PLANET NOW

PLANET NOW

EFFECTIVE STRATEGIES FOR COMMUNICATING ABOUT THE ENVIRONMENT

JESSICA REID

NEW DEGREE PRESS

PLANET NOW

Effective Strategies for Communicating about the Environment

ISBN

978-1-64137-942-7 *Paperback*
978-1-64137-748-5 *Kindle Ebook*
978-1-64137-749-2 *Digital Ebook*

To Earth and the life that depends upon it, may we protect and honor you.

CONTENTS

———

INTRODUCTION

———

Across the United States in the mid-twentieth century, human impact created pollution in the air and water, destroying human health, killing wildlife, and decimating ecosystems. Chemical companies had taken over as powerful industries. Biologist and writer Rachel Carson made a significant mark when she wrote about this problem in *Silent Spring* in 1962. The book became successful and was an impetus for the environmental movement in the United States. It eventually led to the first Earth Day and the creation of the Environmental Protection Agency in 1970. Through the power of words, people began to understand the importance of recognizing human effects on the environment. Although Carson's writing sparked much controversy, especially from chemical companies, it eventually led to the action necessary to create and maintain a cleaner environment.[1]

Today, environmental challenges like climate change, pollution, and waste reduction continue to grow more complicated and momentous in many interconnected ways.

———

[1]   Eliza Griswold, "How 'Silent Spring' Ignited the Environmental Movement," *The New York Times*, September 21, 2012.

We hear about climate change all the time. "Global warming" used to be the common name for the changes affecting many earth systems, and the earth is warming overall. In 2019, the average global temperature was more than 1 degree Celsius and about 2 degrees Fahrenheit higher than in 1880.[2] And while the earth's temperature is indeed rising overall, warming is only part of the threat facing our planet. The climate in many areas of our world is changing in numerous ways and will change even more dramatically over the next few decades. While a temperature jump of one or two degrees may appear small, consider this fact: tree rings, ice cores, and coral reefs show that the earth's average temperature tends to be stable over time, and small temperature changes can greatly impact the environment.[3] As a kid in the 2000s, I found that the pervasive question seemed to be: Is climate change real?

Recently, we have seen a spate of record-hot years; nine of the ten warmest years on record have occurred since 2005, with the hottest five years occurring between 2015 and 2019.[4] In the heat of these emerging moments in climate history, a movie did what no report, study, or press conference had been able to accomplish—that is, planting an awareness of climate change into the public's mind.

The year was 2006. The film was former Vice President Al Gore's iconic documentary *An Inconvenient Truth*.[5] A few

2   "World of Change: Global Temperatures," NASA Earth Observatory, accessed July 8, 2020.

3   "Climate Change: How Do We Know?" NASA, accessed July 8, 2020.

4   "2019 was 2nd hottest year on record for Earth say NOAA, NASA," NOAA, January 15, 2020.

5   Al Gore et al, *An Inconvenient Truth*, Hollywood, CA: Paramount, 2006, 96 minute.

years later, I was a middle school student whose worldview would be transformed by that truth. I remember feeling a sense of urgency and obligation to do something to stop this catastrophe from happening.

But while the movie's impact on public awareness of climate change is undeniable, the media's politicization of the film fractured US public opinion into two opposing camps: the so-called "liberal elitists" and "climate deniers."

Sociologist Robert Brulle has studied the opposing sides, a trend he has labeled the "climate change countermovement." In his research with Drexel University, Brulle identified "well-organized" groups, like think tanks, that oppose climate action. As they have recently had more trouble proving climate change nonexistent, they've turned to other arguments, such as saying that mitigating climate change is too costly for the economy.[6] However, the costs of inaction are much greater than the costs of preventing climate change in the present.

Americans have spent too long arguing about whether the issue is real. Waiting past 2030 to take climate action would raise the costs of decarbonization by 50 percent. Some people may say we should wait for more technology, but human environmental impact is continuing to grow, and society must invest in and research new technology now so it's more affordable in the future.[7]

---

6   Jason M. Breslow, "Robert Brulle: Inside the Climate Change 'Countermovement,'" Frontline, PBS, October 23, 2012.

7   Ilario D'Amato, "Delaying Climate Action Will Raise Costs 50 Percent: World Bank Report," The Climate Group, 12 May 2015.

To solve the problem, we must understand climate change and its effects.

Climate change occurs when certain gases, like carbon dioxide and methane, are released into the atmosphere from many sources, such as power plants. They trap heat from the sun and warm the earth's atmosphere. While these gases have been around since long before humans, they've been released in high quantities following the Industrial Revolution of the 1800s. These gases are called greenhouse gases because they keep the earth warm, like a greenhouse.

Along with seventeen other domestic and international associations, the International Panel on Climate Change (IPCC) has determined that climate change is real. Made up of scientists across many disciplines, this group is the United Nations' panel that reviews thousands of research papers per year to determine the science of climate change and make policy recommendations. This warming causes many effects, such as increased or decreased precipitation and changing temperatures in most regions of the world.

An IPCC 2019 special report explains the impacts of the earth warming more than 1.5 degrees Celsius above preindustrial levels. The report says global warming of 1.5 degrees Celsius will likely occur between 2030 and 2052 if emissions continue at their current rate. Further, it warns, "Climate-related risks to health, livelihoods, food security, water supply, human security, and economic growth are projected to increase with global warming of 1.5 degrees Celsius and increase further with 2 degrees Celsius."[8] Keep in mind that emissions and warming will continue increasing beyond this point if we continue on the path we are on now.

We do not have long to mitigate irreversible future consequences like extreme weather, sea level rise, and food and water scarcity—issues that are already starting to present themselves. Although directly linking specific weather events to climate change is challenging, the preponderance of devastating events like the fires in Australia, California, and the Amazon Rainforest, as well as hurricanes like Harvey and Maria, exemplifies some of the disasters that can result from climate change. Increasing hot and dry conditions have led to an increase in wildfires. Warmer nights have prevented fires from dying down, making them worse.[9] Yale Climate Connections attributes more intensive hurricanes to warmer ocean water, cautioning that "as the world warms, we're setting the stage for severe hurricanes that could do even more damage than what we've seen before."[10]

"Since 1975, there has been a substantial and observable regional and global increase in the proportion of category four to five hurricanes of 25-30 percent per degree Celsius of anthropogenic global warming," a *Climate Dynamics* study

---

8   "Global warming of 1.5 °C," Intergovernmental Panel on Climate Change, 2019, 4-9.

9   Jessica Merzdorf, "A Drier Future Sets the Stage for More Wildfires," NASA, July 9, 2019.

10 Sara Peach, "Sea Surface Temperatures Drive Hurricane Strength," Yale Climate Connections, August 3, 2016.

concludes. Therefore, hurricanes are indeed becoming stronger as humans cause climate change.[11]

NASA says the sea level has risen about eight inches in the past one hundred years.[12] Cities like New Orleans and Miami already face rising waters and could be underwater by 2100.[13]

Yale Climate Communications says food scarcity is also already occurring due to climate change. For example, California lost about $3.8 billion in agriculture from 2014 to 2016 because of drought.[14]

These terrible events are portents for worse results of climate change in the coming years.

Further, over a million plants and animals could go extinct within decades. The deterioration of habitats and living conditions can significantly reduce biodiversity.[15]

Nevertheless, people still have differing opinions about climate change and can ridicule those with other perspectives. The expressed differences lend to demeaning statements about

---

11   Greg Holland and Cindy L. Bruyère, "Recent intense hurricane response to global climate change," *Climate Dynamics* 42 (2014): 625.

12   "Climate Change: How Do We Know?" NASA, accessed July 8, 2020.

13   Aria Bendix, "8 American cities that could disappear by 2100," *Business Insider,* March 17, 2020.

14   Daisy Simmons, "A brief guide to the impacts of climate change on food production," Yale Climate Connections, September 18, 2019.

15   "UN Report: Nature's Dangerous Decline 'Unprecedented'; Species Extinction Rates 'Accelerating,'" United Nations, May 6, 2019.

opponents. Here we will explore fresh and convincing ways to communicate.

Compounding upon climate change and closely related to this problem are the issues of pollution and waste. The materials system is not sustainable, with the United States producing more than 230 pounds of plastic waste per person each year.[16] About eight million tons of plastic worldwide are being thrown into the ocean every year, and at this rate the oceans will have more plastic than fish by 2050.[17] Further, 91 percent of plastic is not recycled.[18] A circular economy focuses on reusing materials rather than constantly producing more and more. This solution can also decrease greenhouse gas emissions by reducing production.

Many scientists and entrepreneurs are coming up with innovative solutions to mitigate climate change and reduce waste, such as renewable energy sources and the circular economy. However, these solutions can only be successful when people know about them and can support large-scale, systematic implementation. Communication is necessary, but it is challenging when the environment is a controversial subject. While some people accept the current views of environmental science and think the need for action is not up for debate, others are unsure if the current findings are valid or what they as individuals can do to make an impact.

16 Emily Holden, "US produces far more waste and recycles far less of it than other developed countries," *The Guardian,* July 3, 2019.

17 Sarah Kaplan, "By 2050, there will be more plastic than fish in the world's oceans, study says," *The Washington Post,* January 20, 2016.

18 Laura Parker, "Here's How Much Plastic Trash Is Littering the Earth," *National Geographic,* December 20, 2018.

So much of today's climate change discussion is unproductive and abstract. Spouting facts is not bringing about the changes we need. The state of this issue is frustrating. I know we have seen some success, and over the past year I've studied what *has* worked to learn how to better drive effective communications.

According to the Yale Program on Climate Change Communication, in 2019:

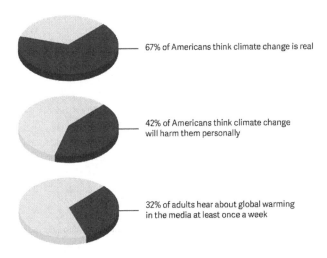

67% of Americans think climate change is real

42% of Americans think climate change will harm them personally

32% of adults hear about global warming in the media at least once a week

Opinions and exposure to climate change facts vary among Americans.[19]

The media landscape has evolved over the past couple decades from widely debating the science of climate change to discussing a hopeless crisis. Political polarization and changes in digital

19 Jennifer Marlon, Peter Howe, Matto Mildenberger, Anthony Leiserowitz, and Xinran Wang, "Yale Climate Opinion Maps 2019," Yale Program on Climate Change Communication, September 17, 2019.

media are altering the ways people communicate about climate. People are communicating across more platforms than ever, from the kitchen table to newspapers to podcasts to social media.

One challenge when discussing climate change is the need to prove it. Those of us who feel climate action is important should try to discuss the proof and credibility of climate change when we can. However, we should focus not on arguing but on communicating feasible solutions to help others understand steps to mitigate environmental issues. Even if people are reluctant to accept current climate science, helping them understand the economic benefits of cost-saving measures such as solar panels or reusable products can encourage them to take more sustainable measures in their own lives.

Personally, I witness environmental impact when I return home from UNC–Chapel Hill to my hometown of Apex, North Carolina, where land once covered by pine trees has become blanketed with single-family homes. Located close to the Research Triangle Park, Apex was named *Money Magazine*'s "Best Place to Live in 2015."[20] Tens of thousands of people have moved to the town in my lifetime. I feel grateful to have grown up in a place where many people want to live, and I know that development is important to the economy. However, extensive land use and clear-cutting make me worried about the future of my hometown. I'm not alone in my concern. Many Apexians would prefer slower growth and less deforestation.

---

20 Rebecca Troyer, "Money Magazine: Apex is the No. 1 place to live in the US," *Triangle Business Journal,* August 17, 2015.

I grew interested in the environment and sustainability from seeing this environmental impact up close. Since I realized this problem is happening in communities around the world, I've become concerned about the future of life on earth. Beyond Apex, I grew up appreciating animals and plants, from my family bringing me to places ranging from the North Carolina Zoo (the world's largest natural habitat zoo) to the Waterman Conservation Education Center in Apalachin, New York. I also love spending time outdoors, like gardening, hiking, and kayaking in the waters of North Carolina's Crystal Coast, close to the areas where Rachel Carson did some of her work.

My concern has led me to make studying and advocating for the environment a daily part of my life. I am majoring in environmental studies at UNC–Chapel Hill, where I am a leader in and run communications for many environmental organizations. I'm also studying public policy and journalism and public relations. I have experience being a policy analysis and communications intern at the EPA's Office of Air Quality Planning and Standards, serving as a communications intern at the East Coast Greenway Alliance in North Carolina, conducting environmental law research at Brenner Law Firm, and sharing stories about sustainability entrepreneurs around the world for the nonprofit Heart of Waraba. My environmental work started out with my collecting bottles to recycle before school in third grade, and I've learned a lot since then about protecting the environment and affecting change on larger scales.

As you'll read about later on, I'm part of an environmental honors society that held a conference on the circular economy to bring together the UNC–Chapel Hill community around this new idea. Since then, I regularly hear students and faculty referencing the circular economy. I find we have fascinating ways to communicate about the environment, like this conference, and lessons that can transfer from mass communication to individual communication.

One example of impactful communication is environmental science student Brooke Bauman creating a podcast about the preponderance of plastic. She used this mode of communication to help people understand her journey to learn about the environmental harm of plastics use as well as alternatives.

Learning about different ways people communicate can help others apply these new and useful methods to their own lives. People need to not only hear about climate science but also listen to this topic so we can achieve progress toward a more sustainable future.

"Sustainability is meeting the needs of the present without compromising the ability of future generations to meet their own needs. It requires the reconciliation of environmental, social equity, and economic demands—also referred to as the 'three pillars' of sustainability," explains UCLA Health.[21]

In this book, I will explore the ways people communicate about these issues, whether as part of their careers or as individuals. Through speaking to different individuals, I will cover tips for mass communications about the environment as well as ways to discuss them with each other. This book will also explore strategies to reach audiences, like connecting

---

21 "What is Sustainability?" UCLA Health, accessed July 8, 2020.

to others' values, using psychology, and thinking about how the Myers-Briggs personality types and enneagrams can be employed to communicate about these issues in ways your audience can understand. Further, I will discuss the importance of intersectionality and inclusivity in environmental communication. At this time, we have technology to help us not suffer the impending consequences of human impact on the environment. However, communication is necessary for all to understand what must happen and to encourage changes to be made at policy levels. To change hearts, minds, and habits, we must reframe our message to show there is hope if we take the necessary action now.

Discussing the environment can be challenging when most of us are not scientists. We might feel intimidated to debate someone about climate science when we don't understand the intricacies ourselves and when exact future predictions about changes are just estimates. Many of us might not feel confident advocating for climate science if we are not experts. But, as science communicators like Susan Joy Hassol and Kate Sheppard will explain later in this book, for climate skeptics, hearing about climate science from people they know and with whom they have relationships can be most effective.

We can start by making individual sustainable choices—a major one being the reduction of our meat consumption. While that step is necessary, we need to do more. Understandably, not everyone has the ability to make huge changes, reach net-zero emissions, or live waste-free. Certainly, some people are just trying to get by, and different people have different abilities. However, if you have the privilege of being able to do something about climate change and you prefer a sustainable world to a much more uncomfortable world, the time to do something about it is now. Even if you are not old enough to vote, you have the power of a youthful voice to share your ideas. If you are able to vote, do your civic duty. I am not personally endorsing any particular candidate or party in this book, and the issue is not Democratic or Republican. The issue is the powerful groups that oppose climate action and politicize and obfuscate climate change science. Most important is how we can all come together to create a better world for ourselves and future generations.

Notably, this book focuses on environmental communications in the United States. Of course, environmental action is needed across the globe, but the United States is a good place to start for those of us who live there, and many of these principles can be applied elsewhere.

If you are a self-described tree-hugger and already passionate about the environment, you will enjoy learning how to express your ideas to others to help them understand the issues we are facing.

Or maybe you know climate change is an issue but don't feel like you know much about the environment. You may want to better understand how to talk about it and how to sound like you know what you're talking about.

If you want to improve your ability to talk about the environment at any scale, read on.

Climate change is an issue not just for the future, but for the present as well. We need to be strategic in how we communicate about it, which is why we need to "planet" now.

PART ONE

HOW WE
GOT HERE

1

WAVES OF CHANGE

Waves begin as small ripples when wind blows against the water. Then the blue-green waves heighten and move energy from one part of the ocean to another or to shore. Sometimes waves can carry energy far from their origin.[22] Nature does not sit still, and neither should humans.

While we need many people to take action to live sustainably, we can only create impactful change by communicating the urgency of environmental action to those in power and to everyone else around us. We can multiply our efforts and create waves of change through communication.

I have long wondered about the best ways to have an impact on the environment when it feels like individual actions do not make a meaningful difference. While we can make an impact in more than just one way, I had an epiphany when I realized words can have exponential effects: communicating thoughts and perspectives can lead to many more people becoming aware of important ideas. Words will not solve

---

22 Jane McGrath, "How Wave Energy Works," How Stuff Works, July 15, 2008.

environmental problems, but they are a chief element for affecting the ways people think as well as the priorities of elected officials and corporate leaders. If the mainstream media and social media did not share and amplify climate science, most people would probably not be aware of the issue, and we would not see as much support for those in power taking action.

Words have the power to educate and inspire people to work toward sustainability.

Awareness of environmental issues comes into play in my own life. Because of the media I regularly consume about the environment, this topic constantly cycles through my mind. I think about waste whenever I'm throwing something away. Media has inspired me to make the effort to compost and recycle when possible, even if doing so is less convenient, and to use reusable products like water bottles and shampoo and conditioner bars. More significantly, media coverage of environmental issues has taught me the need to work for larger environmental change.

"I must believe communication has a tangible impact because I'm engaged daily in environmental communication, and yet it's not clear that it has a sufficient impact. There are lots of examples of news stories about pollution [that force] government and business to clean up pollution, and we've seen some progress on that front," said Jeff McMahon.

McMahon is an energy and environment journalist at *Forbes*, as well as a professor at the University of Chicago. He started writing about energy and the environment when he learned his college was getting rid of radioactive waste in a dumpster. He teaches journalism, argument, and scientific writing.[23]

"I think we would not have a Paris Agreement," he said, "if the gravity of the climate crisis were not communicated to people globally."

Even so, he isn't convinced that the Paris Agreement, wherein many countries around the world agreed to reduce their contributions to climate change, will have a sufficient impact.

"Communications like Rachel Carson's *Silent Spring* and Al Gore's *Earth in the Balance* and *[An] Inconvenient Truth* helped mobilize the environmental movement and then the climate movement, but environmental damage has continued to worsen during the lifetime of these texts," continued McMahon.

McMahon is not alone in believing in the power of communication. Maggie Kash believes communication can absolutely make a tangible impact on the environment. Kash, who serves as director of communications at the Sierra Club, said, "Communications strategy is essential to moving policy, and that is the case for environmental policy, especially since communications is tied to public opinion and how policymakers make their decisions."

"In terms of advocacy, if you're not communicating about what you're advocating for, then you're not doing the work,"

---

23   "Jeff McMahon," *Forbes*, accessed May 18, 2020.

continued Kash, who spoke to me as a communications professional and not as a representative of her employer.

In order to mobilize people, the spread of information is necessary. Still, action must follow communication, and strategy can lead to intended changes.

Kash discussed a specific instance when communications strategy led to a tangible impact on environmental goals. Former EPA Administrator Scott Pruitt is known for making decisions that did not advance the EPA's mission to protect public health and the environment. The Sierra Club's communications strategy was responsible in large part for the removal of Scott Pruitt from the position of EPA administrator. Kash explained, "He had a lot of objectionable practices and had close ties with polluters. As EPA administrator, he was doing all sorts of shady things and corrupt dealings."

Kash said the Sierra Club had a legal and communications strategy from the beginning that included requesting emails from the EPA under the Freedom of Information Act. The goal was to find evidence of "wrongdoing and corruption" in order to publicize Pruitt's actions. The Sierra Club also used tactics like online campaigns and ads on President Donald Trump's favorite television shows to influence this change and force Pruitt to resign from his position.[24]

Talking about the environment is not enough, but materials like *Silent Spring* and *An Inconvenient Truth* have had

---

24 Rebecca Leber, "Scott Pruitt's 'Dirty Dealings' Stir a Campaign to Oust Him From the EPA," *Mother Jones*, accessed March 28, 2018.

effects. As mentioned previously, *Silent Spring* led to major environmental action, and *An Inconvenient Truth* increased the public's understanding of climate science. Seeing the movie's graphs of changes in the earth's atmosphere compared to human emissions helped me understand the rapid and extreme problem that climate change poses.

Youth activist Greta Thunberg has brought people together from around the world to strike for climate and stimulate news coverage. She boldly tells the world, especially politicians, that we need to take climate action immediately. Her activism has truly caused waves of change by galvanizing other young people to be environmental activists too.

"I think Greta Thunberg's messaging is crucial to properly orienting our response to the climate crisis, but will it be effective? It's too early to say. Where are the results? Certainly environmental communication has been important to raising our species' awareness of the damage it is doing, and to mobilizing a response, but so far, awareness hasn't been enough and our response hasn't been adequate," said McMahon.

While Thunberg's message is inspiring, it will not compel everyone to make changes, particularly because people may feel that making big changes quickly is quite radical.

Framing is useful for encouraging someone to receive a message, because a person may need to feel like they understand a viewpoint to want to learn more. It can also encourage action. Communicators should think about how to best reach different audiences. Connecting with audiences is a topic that will be detailed in depth later in this book. Framing climate change as an issue that individuals should tackle, like using metal straws, will not solve the crisis. On that note, calling climate change a crisis might discourage people from listening out of fear of change. Thinking about how environmental issues are and should be framed is important. A later chapter will discuss how framing successfully reduced food waste in San Francisco.

Here are some specific ideal ways to frame environmental issues:

- We should frame the need to help the environment as something necessary for humans to survive and thrive, not just as something to help nature. Like many people, I care about preserving natural landscapes and biodiversity, and I want to protect endangered species. But not everyone thinks the environment is a priority or a reason to reduce profit.

- We should frame climate change from an environmental justice perspective. This issue is not just something for the privileged to tackle, and significant environmental action cannot occur without including all people in conversations about solutions. Climate action is a way to empower communities, since communities that are already marginalized will often face the worst results of climate change.

- We should frame climate action as something that can benefit rather than harm the economy. This idea applies both in the near future by creating renewable energy jobs and in the coming decades by mitigating the money we will have to spend adapting to the terrible effects of climate change like natural disasters and agriculture loss.

- We should frame environmental problems as being in our control. The future is not hopeless, but we must urgently take action beyond individual choices that may make people feel good but not help much. We should avoid acting like the world is ending in the near future. The earth is not going to disappear in a couple decades, but we will see—and already are seeing—devastating consequences for life.

By using strategic framing, we can more effectively convey the importance of helping the environment and make it relevant to all.

Communication alone is not enough to change the world, but it is necessary. My hope is that this book will help readers gain knowledge and tools that will benefit them when communicating to spread environmental awareness and create change. Communicating can help social leaders understand the need or feel pressure to make decisions in the best interest of the environment, which is also in the best interest for humanity.

"Communication has the ability to help people feel empowered and feel like they have the information they need and feel like they can make decisions and take action in their daily lives, and that can change what happens [at your local, state, and national levels]. Ultimately, we elect the people who make the big decisions," said Kate Sheppard, *HuffPost* senior enterprise editor and teaching associate professor at UNC–Chapel Hill's Hussman School of Journalism and Media.

The need to make large changes does not mean individual changes do not matter. They can motivate people to contribute to larger change.

"I am very much a proponent of: yes, it's good to do what you can as an individual, but individual choices are not going to be the biggest, most important changes. They can help us feel empowered and they can raise awareness, but honestly, the biggest changes are at the national and international political level and economic level," continued Sheppard.

"But it's helpful to tell people that they can do things personally, because that helps them feel like they're engaged. People's empowerment and people's awareness trickles up to your leaders. You vote for these leaders.... I have people asking me what [they can] do, and I am like, 'Honestly, move to a place that needs the change and vote there,' because we tend to be pocketed in this country in different areas where everybody agrees that we should act here and everybody doesn't agree here," explained Sheppard.

Therefore, we can take steps to reduce climate change by helping individuals understand what they can do but also encouraging people to support more impactful policy changes.

I talked with Susan Joy Hassol, a well-known expert in effective climate science communication. "I think public understanding is important because it can lead to people taking action. If people don't understand the problem, if they don't understand what's causing it, and how serious it is, and how urgent it is, then they're not likely to do anything," said Hassol.

Even when some personal actions seem insignificant, we as individuals can take certain actions that are impactful in our communities and beyond. Of course, people can and should make sustainable choices about what they use, eat, and drive.

But Hassol offered many great suggestions for more significant personal actions, like "taking part in a climate strike or a rally, or organizing your neighbors to sue a utility that's operating a power plant that's poisoning the community. Calling your representatives and senators and mayors are also personal actions, [as well as] telling them that you care about climate change, that it's a voting issue for you. Voting is a personal action. When we choose candidates, we should look at what they think and do and what their policies are with regard to climate change. And if their policies aren't strong enough, demand better from them. And when they're in office, hold them accountable."

Sometimes I hear people complain that others talk too much and don't act or do enough, but talking about actions is a true necessity and the first step for creating change.

"When people start taking those voluntary actions, it can make them want to do more. And it can make them want to work for systemic change. And it's that systemic change that we need to really solve the problem," continued Hassol, echoing Sheppard's point that people should still take personal action to feel empowered.

Anxiety about the future state of the environment is an issue many people, especially youth, face these days. However, taking individual action can help mitigate anxiety and encourage people to continue making adjustments to their daily lives that keep them motivated and make an impact.

"People have to believe that what they do matters, that what they do will make a difference, that it's not just a drop in the bucket. The small personal actions can seem like a drop in the bucket, but the larger actions that we take with others have more efficacy," said Hassol. For example, advocating for policy requiring more renewable energy can lead to progress on a large scale.

Change will come from acting and not just talking, but effective communications is a necessary starting point for drawing people into supporting changes toward sustainability.

Yet the exact changes that need to happen can be confusing and overwhelming with many local, state, and federal policy proposals and uncertain science, such as the precise effects of climate change or amount of fossil fuels left.

What is the message we should be communicating?

Hassol expressed the importance of having a "simple, clear message" that we need to keep fossil fuels in the ground and move toward clean energy. We need to increase energy efficiency and electrify everything, including transportation. As we electrify everything, we must shift to renewables, like solar, wind, and hydropower for electricity rather than coal and gas.

To get there, we will need policies, all with pros and cons. We must educate ourselves on policy options. Most importantly, we must keep moving in the right direction toward sustainability.

"The policy options include the Green New Deal, putting a price on carbon, [and] a revenue neutral carbon tax as Citizens' Climate Lobby proposes and supports. Any of those things will move us in the right direction. And we know what we need to do with regard to land use: stop cutting down trees and plant more trees, and support agriculture that is regenerative and keeps carbon in the soil. But the biggest thing we have to do is keep the fossil fuels in the ground. Keep the coal, oil, and gas in the ground, stop subsidizing and supporting any new fossil fuel infrastructure: no new wells, no new mines, no new pipelines. We need to phase out fossil fuels as quickly as possible. So really, when it comes down to it, it's pretty simple. Those are the things we need to do, and any policies that move us in that direction are good things to support," Hassol expounded.

Put that way, we clearly have core steps to take for sustainability, and these are the points we should emphasize when encouraging environmental action.

Communication is impactful for policymakers and politicians. While they do not always follow public opinion, they are more likely to adhere to their constituents' desires if a policy issue is salient. Therefore, communication can build the policy agenda and lead to tangible change. Constituents should express satisfaction with decisions they support to incentivize lawmakers to continue listening to them.[25]

The decisions of people in power do rely on the will of the public, and informing the public about environmental issues can lead to pressure for world and community leaders to make sustainable choices. Tracing the direct impacts of what we

---

25 Christopher Wlezien and Stuart N. Soroka, "Public Opinion and Public Policy," *Oxford Research Encyclopedias* (April 2016).

communicate can be difficult, but strategic communication does lead to success, creating a ripple effect as people share their knowledge and thoughts with others. A new tide is coming in as we fight to save our only home before it is too late. Together, we can create synergy to bring a more sustainable and just future. Effective environmental communications creates waves of change.

Next, we will discuss how the current landscape of a polarized and digital world affects environmental communications today.

DISCUSSION QUESTIONS:
- After reading this chapter, how would you explain the necessity of strategic communication for advocating for environmental change?

- Have you ever advocated for change in your community or beyond? Did you find some strategies successful, and if so, what were they? What challenges did you encounter?

- Have you ever changed your own opinion or taken action related to an issue? If so, what caused that change or what made you take action?

2

POLARIZED SIDES

The statistics are staggering:

People who say "protecting the environment should be a top priority for the president and Congress."

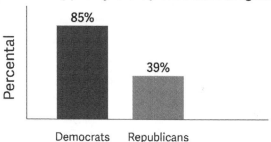

Eighty-five percent of people who are or lean Democratic and 39 percent of people who are or lean Republican say that "protecting the environment should be a top priority for the president and Congress." Similarly, 90 percent of people who are or lean Democratic and 39 percent of people who are or lean Republican "think the federal government is doing too little to reduce the effects of climate change," according to a 2019 Pew Research Center study.[26]

---

26 Brian Kennedy and Courtney Johnson, "More Americans see climate change as a priority, but Democrats are much more concerned than Republicans," Pew Research Center, February 28, 2020.

The concern is not just that some people do or don't believe in climate action, but that this divergence in opinion relates to party lines. Thinking of ourselves in groups makes reaching people with differing opinions more difficult.

Tribalism is the idea that people tend to stay loyal to a group, which can create animosity among groups that differ from one another.[27]

My gears started turning about polarization and its effects on the environment when I took an honors seminar called "Darwin's Dangerous Idea." In this class, I had a group project with the objective of exploring and proposing how evolution might have caused a social phenomenon—in my case, polarization. I developed an understanding of polarization and some of its effects. From this project, I learned that people tend to strongly identify with their beliefs and may not be willing to compromise. Polarization allows people to band in groups against enemies, which could have been beneficial for survival in the past. Today, it prevents our society from solving climate change, which could lead to the demise of our species and other species.

Polarization has increased in the United States since the 1980s, when moderates started disappearing.[28] The increase in ideological polarization has been greatest among the most politically engaged. In 2017, 43 percent of people who are or lean Republican reported they identify that way because they

---

27   George Packer, "A New Report Offers Insights Into Tribalism in the Age of Trump," *The New Yorker*, October 31, 2018.

28   Drew DeSilver, "The polarized Congress of today has its roots in the 1970s," Pew Research Center, June 12, 2014.

are "against what the Democratic Party represents," and 40 percent of people who are or lean Democratic reported they identify that way because they are "against what the Republican Party represents."[29]

Polarization applies to many aspects of life. While some might argue that opposites attract in romantic relationships, the reality is that in most marriages, both partners are together on the political spectrum, whatever their ideology. Twenty-five percent of couples surveyed consisted of two Democratic partners, 15 percent had two partners who identified as independent, and 30 percent had two Republican partners, according to a 2016 study conducted by Eitan Hersh, a Yale University political science professor, and Yair Ghitza, the chief scientist at the political data firm Catalist.[30]

A factor that contributes to polarization is geographic isolation. People with a more liberal ideology are moving to urban areas, and people who are more conservative are moving to rural areas.[31] People with conservative beliefs may not be exposed to as much conversation about the environment when in rural areas, although many rural areas are going to experience the negative effects of climate change, like reduced crop yields. As issues grow and media attention on those issues increases, people may perceive these ideas as

29 "Most who identify as Republicans and Democrats view their party connection in positive terms; partisan leaners more likely to cite negative partisanship," Pew Research Center, October 4, 2017.

30 Eitan Hersh, "How many Republicans marry Democrats?" *FiveThirtyEight*, June 28, 2016.

31 Aaron Blake, "People move to places that fit their politics. And it's helping Republicans," *The Washington Post*, June 13, 2014.

threats and grasp tighter to those with similar views. This phenomenon can lead to polarization and the need to fight against the "other side."

When people congregate in like-minded areas, they can have greater difficulty understanding other viewpoints; people who are liberal may not understand how to reach people with conservative views without acting like conservatives are close-minded or uneducated, which are negative stereotypes. Such stereotypes can push people away and cause conservatives to distrust liberals. We should not make the environment more partisan than it already is these days but rather be inclusive in the fight for the future.

Identity also can lead to polarization. Each party has its distinct characteristics that it feels defines its members, as well as characteristics members of these parties commonly assign to other parties. Notably, both Democrats and Republicans view each other as close-minded. Again, negative views of the people making up both parties can contribute to people not working together.[32]

This polarization can be problematic because people may not support certain issues solely because of their party. However, while the Democratic Party is generally considered the more pro-environment party, that does not mean all Republicans are opposed to environmental and climate action, especially those who are young. A 2019 poll from Ipsos and Newsy showed that 77 percent of millennial and Gen Z Republicans viewed climate change as a serious threat. I have friends and family

---

32   "How partisans view each other," Pew Research Center, October 10, 2019.

who are simultaneously Republicans and environmentalists. Nevertheless, we need to help people understand and be open to learning about the environment so people can be less biased about these issues solely from taking political sides.[33]

"The causes of America's resurgent tribalism are many," wrote Amy Chua and Jed Rubenfeld, both professors at Yale Law School. "They include seismic demographic change, which has led to predictions that whites will lose their majority status within a few decades; declining social mobility and a growing class divide; and media that reward expressions of outrage."[34]

Political polarization ebbs and flows between the extremes across time. What's the state of political tribalism in the Donald Trump era?

A study out of the Annenberg School for Communication at the University of Pennsylvania found no increase in polarization from 2014 to 2017. Professor Yphtach Lelkes, who conducted the study, said, "I've been studying polarization for a long time, and elite discourse is arguably at its worst, which led us to theorize that partisanship would be worse since Trump took office. But we found that things really have not budged."[35]

---

33 Kate Yoder, "On climate change, younger Republicans now sound like Democrats," *Grist*, September 9, 2019.

34 James Fallows, "A Nation of Tribes, and Members of the Tribe," *The Atlantic*, November 4, 2017.

35 "Donald Trump's Election Did Not Increase Political Polarization," Annenberg School for Communication at the University of Pennsylvania, October 11, 2019.

Tribalism is inherent in society. However, people from different sides working together is not impossible. Republicans and Democrats, fiscal conservatives and progressive liberals, could and should speak to and learn from one another about specific challenges related to the environment.

The viewpoint that one side is wrong often leads to neither side being willing to work with the other. The environment should not be about party but about the need to act sustainably for a better future for most or all people. The solution is to find ways to explain these issues at individual and larger scales to better reach people.

We need improved communication about the environment because media is convoluted with such brief information and misinformation that the issue is difficult to understand. In a polarized political era, people tend to choose a side and stick with it, even if they do not know the facts and learn more. The environment is a scientific and social issue made political because action to mitigate and prevent environmental problems requires political action.

Bipartisanship has worked in some instances, like with the 1973 Endangered Species Act. Senator Harrison Williams and Representative John Dingell, both Democrats, wrote the bill after Republican President Richard Nixon expressed the need to protect endangered species. This act led to 99 percent of listed species avoiding extinction.[36]

---

36 "The US Endangered Species Act," World Wildlife Fund, accessed July 8, 2020.

Another successful example of people coming together to help the environment was in the same time period of the Endangered Species Act. Many people rallied for change after Rachel Carson wrote about the effects of chemicals on the natural world and human health. She studied nature and the adverse effects humans and industrialization were having on the environment and then shared her findings with the public, which led to greater environmental awareness. More recently, members of both parties came together to pass the Twenty-first Century Cures Act in 2016.[37]

Beyond getting political parties to come together within the United States, countries have come together to agree to global change. However, their measures have proven ineffective. The Kyoto Protocol is a well-known international agreement to reduce greenhouse gas emissions and slow climate change. This agreement was signed in 1997 and took effect in 2005, but the world today remains reliant on fossil fuels. The agreement did not impose severe consequences on countries that failed to meet the standards of the accord, so participants had no sufficient incentives or repercussions to comply. Considering the failure of these policies, noteworthy change requires the public to be knowledgeable about climate change, according to Paul Mayewski, director of the Climate Change Institute at the University of Maine Orono.[38] The public should and must be passionate about environmental action.

37   Robert Pear, "Cures Act Gains Bipartisan Support That Eluded Obama Health Care Law," *The New York Times,* December 8, 2016.

38   Laura Poppick, "Twelve Years Ago, the Kyoto Protocol Set the Stage for Global Climate Change Policy," *Smithsonian Magazine*, February 17, 2017.

Not all international environmental agreements are destined to fail. The Montreal Protocol was another international environmental effort. In the late 1980s, nations agreed to reduce use of the chemicals destroying the ozone layer of the atmosphere. Today, the ozone layer is in better shape. Challenges arise when the science becomes political and the media obfuscates the issue.[39] We must find ways to overcome these problems and stay inspired to take action.

Rachel Carson was insightful in finding the problems for humans and then getting that information out to the public. That's just one of the many reasons I've come to hold her in high regard. She spread ideas through her writing and was editor-in-chief of the US Fish and Wildlife publication. Carson was curious and dedicated to what she loved. She found her passion and changed the world with it. In particular, she was concerned about the widespread use of the pesticide dichlorodiphenyltrichloroethane (usually called DDT), which killed fish and birds, hence the name of her most well-known book, *Silent Spring*. She was relentless in working for what she believed in.[40]

Carson did receive much backlash from the agriculture industry, including attacks on her claims and character, but that did not stop the general public and lawmakers from taking action. As I mentioned, her work prompted a public call to action; President Nixon formed the EPA, Congress passed the Clean Air Act and Clean Water Act, and Earth Day became a holiday.[41]

---

39   Ibid.

40   Eliza Griswold, "How 'Silent Spring' Ignited the Environmental Movement," *The New York Times*, September 21, 2012.

41   Ibid.

Industries' attacks on environmental science were just beginning, so much less polarization and controversy around the environment existed at the time.[42]

"No single work has had the impact of *Silent Spring*," said journalist Eliza Griswold, a 2019 Pulitzer Prize winner.[43,44]

One of my favorite vacation spots is the serene waters of the Bogue Sound in North Carolina, very close to the Rachel Carson Reserve. I enjoyed touring the waters one summer, and I remember seeing wild horses in the distance. I'm thankful that Carson worked to preserve a healthy environment for generations after her, but we still have much work to do.

Her work has inspired me to combine my writing and passion for the environment to persuade others that we must be conscious of our effects on the natural world. Public support is necessary to make significant progress to help the environment, whether by banning chemicals, utilizing sustainable development practices, or implementing other solutions to environmental challenges.

We need to share with others information about potential environmental problems, no matter their political beliefs, so they can understand the challenges we face in the next few decades—challenges I believe are surmountable with quick action, innovation, and commitment.

---

42   Ibid.

43   "Eliza Griswold," *The New Yorker,* accessed July 8, 2020.

44   Ibid.

The Clean Tech Summit hosted by UNC–Chapel Hill illustrates a modern example of people from different backgrounds, like professionals from various industries, college students, and even politicians—both Democrats and Republicans—coming together to discuss innovations in sustainable technology. The opportunity to meet other students and professionals across environmental fields has drawn me to the summit twice so far during my college career. Professionals shared the latest trends and innovations in areas from environmental journalism to the circular economy to ocean technology.

Dr. Greg Gangi led development of the summit and was my professor for the class "Introduction to Environment and Society." I talked to him to learn about the successes of the summit as well as his thoughts about successful communication about the environment, especially climate change.

He told me about the many learning opportunities at the summit. "There's the built environment and then there's the natural environment. And [when] we often think about the environment, we have a tendency in this country not to think about the built environment. Quality of life is largely affected by the quality of our built environment." As people gather to find ways to create a more sustainable built environment, they have a focus on decarbonizing the economy through transportation and electricity innovation at the summit. Entrepreneurship is another important theme, and many professionals are CEOs and company founders.

Gangi said those in attendance have always been mostly environmental and business students, but a more diverse group of students is starting to attend. For example, computer science and entrepreneurship students outside the business school have started to recognize the networking benefits of the event.

One of the struggles I face in wanting to spread environmental awareness is that the people most inclined to listen are those who already care about the environment. Of course, people across many fields can care about the environment, but creating a diverse assortment of topics that connect to the environment can encourage more people to participate in events that spread information about the environment.

Gangi said that to attract more people to the event, "we've had a wider definition of clean energy that's included nuclear. I'm agnostic on nuclear, but there's a lot of the modular nuclear reactor companies that have a footprint in North Carolina. So we have one speaker this year. Last year, we had a panel because of their major presence in the region. We want to include more entrepreneurs and SMEs [small and mid-sized enterprises], so I think it's important to include them. A lot of these are bigger than entrepreneurs, but some of the companies are not huge. The main thing is it doesn't pay to argue about the best way to decarbonize our energy systems; the market will decide which are the best ways. The most important thing is to get on with the job."

As controversial as the environment can get, people like Gangi are trying to help people of different backgrounds understand environmental issues and make appropriate choices. "One of my objectives is to not make it a liberal celebration but to make it very welcoming and bipartisan. I think the only way in a two-party system that change will happen is if you can bridge the partisan divide, so having Republicans there is even more important than the Democrats."

He points to North Carolina's General Assembly, currently Republican-led, as a success story. "North Carolina's been a leader in solar and here's a very strong core of Republican leaders. I can't take credit for that, but I try to nurture that process, but others have really worked on it. And in some ways, you create an industrial complex that was hard to turn back because of its economic impact on rural areas. It's hard to be against something that's helping your constituents, especially if you're in a rural area where there aren't a lot of economic opportunities, where local governments are having problems paying the bills, and all of a sudden something pops up that's generating more than a million dollars in tax base every year.... Not everybody in the Republican Party is gung-ho about renewable energy, but what we've been seeing [the past] seven years in North Carolina is that this group is getting bigger."

Gangi sees the changes as imperative. "These technologies are becoming cheaper and cheaper already. Coal is an economic no-go. But even with things that don't make a lot of economic sense when it comes to investing in new pieces of infrastructure, leaving stranded assets is still something that's very upsetting to owners of those assets. So that's one hurdle. But I think the main thing is [to] not use tribal language. It's to really find ways that can make this appealing across partisan lines." By helping people from multiple sides agree on a common goal, we can create the necessary changes.

Many decisions come down to money. In the case of climate, that can be a good thing. A misconception is that sustainable decisions are not economically viable and that people must sacrifice the economy for the environment. In reality, the two can go together. Taking that route as opposed to debating about the ethical or moral ways to treat nature is often more successful. Gangi said, "I think there's a growing tendency on the left to try to take the moral high ground and [make] a moral argument about our duties to maintain the environment. That's not going to sell because people on the other side don't trust environmentalists. It's more important for the sake of the future to shape your argument in a way that's more convincing, like national security jobs, stronger economy, or what's going to resonate across society, so people should broaden their rhetoric and how they frame things. The North Carolina Renewable Energy Association has been really effective. Groups with a foot in the business community tended to be more effective than groups purely in the environmental space."

Just like in the United States, Australia faces climate disasters and a reluctance to accept climate science. A lack of scientific evidence is not the problem for tackling the crisis in Australia, especially since the fires offer a terrifying visual of the natural disasters that result from climate change. However, the fires did not significantly change government response or cause a public call to action regarding climate change. Political scientist Robert Keohane suggested a "climate-industrial complex" could help achieve political success with climate action. Industries with the power to encourage climate action and a focus on the economy may be critical for taking the necessary steps to deal with climate change.[45]

This "climate-industrial complex" is similar to the increase of renewable energy and the rollbacks in North Carolina, because it may be a way to surpass tribalism and work toward clean energy.

"I know some Republicans who do the right thing but not for climate change reasons, who vote for clean energy because it's a good rural development tool. It employs veterans. It decreases our reliance on foreign energy. In the end, it's not about getting people to see the world either way, but moving the needle forward in the right direction," Gangi continued.

Effective communication that focuses on engaging a broad and diverse audience can help grow awareness and inspire action in a variety of lives and organizations.

The goal may not be convincing the "other side" about a new point of view, but rather establishing common ground that can help us circumvent polarization. More than 75 percent of Americans believe Americans can work together despite polarization.[46] The following chapters will explain more about surpassing tribalism. Then we can work toward mitigating the growing issues that will ultimately affect everyone, like climate change and the many issues accompanying it, from melting sea ice to natural disasters and agricultural loss.

---

45   Paul Krugman, "Paul Krugman: Australia shows us the road to hell," *The Salt Lake Tribune*, January 10, 2020.

46   "The Hidden Tribes of America," Hidden Tribes, October 2018.

DISCUSSION QUESTIONS:

- Recall one way that polarization makes it difficult for people across political parties to join forces to solve environmental problems. Then share an experience you have had when polarization has affected your taking action on an issue you care about.

- How do you predict people can overcome tribalism to make the necessary individual and policy changes that must happen to mitigate the consequences of environmental problems like climate change?

3

THE DIGITAL WORLD

———

You can go just about anywhere and see people staring at a box in their hand. Sometimes people are condescending toward others and society in general for being obsessed with phones and social media. Despite the challenges modern communications methods pose, they also have benefits.

Given the importance of the digital world for sharing knowledge and shaping people's views, we must use digital technology like social media effectively. In 2020, more than 70 percent of people in the United States use social media.[47] The average user in North America spends more than two hours a day on social media.[48] Social media communications can refer to individuals or organizations promoting their own content, others sharing that content, or paid posts. In all forms of social media communications, if you want to make a difference, you should employ certain best strategies.

---

47 "Social Media Fact Sheet," Pew Research Center, June 12, 2019.

48 "Average Time Spent on Social Media (Latest 2020 Data)," Broadband Search, accessed June 14, 2020.

But first, let's consider: Is social media good for environmental communications?

After the deaths of Ahmaud Arbery, Breonna Taylor, George Floyd, and many other Black Americans, the world saw a social media explosion. Everyone seemed to be discussing racial issues from their personal opinions (positive and negative) to resources for learning more or making changes. I had never seen so many people express support for a cause at once. Racial justice and climate justice are closely tied (you can learn more about that in a later chapter), and use of social media can help spread knowledge about the many issues facing our society, as well as the actions we need to take. I myself have come to better understand the connection of racial justice and climate justice from what I see online. It's also amazing how we can use social media to reach and learn from people across the world. People with similar goals can connect. However, challenges also arise with social media. The efficacy of social media in solving giant social issues is unclear, as is how it may hinder communication and action.

Greg Gangi, a professor at the UNC–Chapel Hill Institute for the Environment, is not a big fan of social media, but he said people use it, so communicators must use it. "The problem with social media is that it's shallow. It does allow for unseen targeting, which can be a double-edged sword. You can micro-target people in a way that's positive or negative. I think it is largely being used in a negative way. People have figured out how to use social media as an effective tool to sell fake news. I know social media, for example, was an important part of the Arab Spring. I'm not sure I've seen great examples in the US of people using it for change. The term 'slacktivism' is probably pretty apt for how environmentalists use social media. For example, when people like a post about how climate change is hurting polar bears, they feel good about themselves and that they've made a statement," Gangi told me.

While social media has potential, Gangi does not feel it is reaching that point in the United States. "I actually think on the whole right now in terms of inspiring people to do what's right for the world, it's a negative force, because it's not bringing people [together]. People spend so much time on it that people don't have time to get together and communicate face-to-face. And I don't think you generate ideas in the same way when you're sitting behind a computer, responding to things versus when you're sitting around with people and the ideas are just flowing more. Also, there's more commitment when you're engaged in ways where you're in the same physical space with people. Again, sometimes that's the best and only way to connect, especially in places where maybe it's dangerous to do that. But I think on US campuses, I saw environmental activism just cratered around 2009. Two things that took off then were the recession and social media," continued Gangi.

Although young people can use social media to promote their voices, it can be deleterious for change because it is also a distraction from environmental work. "At some point, [social media is] worse than just being a time suck. Right now I think it's even a threat to democracy. So, for me, I'm waiting for someone to convince me that it can be largely positive."

However, social media does not seem to be leaving. We must respond to the current state of social media because it is a part of life, just like compounding environmental problems.

Allie Omens, a 2020 UNC–Chapel Hill graduate who studied public policy and the environment, also expressed some of the issues with social media hurting activism. "It's so easy for people to just share stuff on Instagram about rainforests and Greta [Thunberg] and all this stuff, but they just click it and share it. It's so self-fulfilling for people, [but you're] not actually making a difference just because you shared a post."

Nevertheless, Omens said she does sometimes see interesting articles about the environment on social media. At times she is glad someone shared an article because she otherwise would not have seen or read it. For example, she recalled seeing an Instagram post that she liked about how we do not need more sustainable fashion brands but rather more structural change. "Any attempt at change is better than nothing; I'm not trying to discount it," said Omens.

Even though social media posts might share knowledge rather than institute change, they can influence some people in ways that can be difficult to track.

Forbes journalist Jeff McMahon gave another perspective on social media: "Digital communication has allowed for a greater proliferation of voices—more people are publishing than ever—but it has also made it easier for people to only listen to those voices they are predisposed to agree with. At its best, it has encouraged a dialogue. Communication from newspapers, television, and radio only went in one direction—from publisher or broadcaster to recipient—but now recipients can respond, and that's potentially a very good thing for knowledge and for accountability. It would be a great improvement to communication if we had somehow preserved the civility and respect for facts that better characterized the analog system." McMahon alludes to the point that some people on social media do not have accurate information. Without a gatekeeper, people may be unable to distinguish the truth.

Additionally, comments are game-changing because if somebody says something untrue, others can offer pushback that might raise more awareness of the truth (though such interactions may be abrasive sometimes). For example, a famous person could post something about the environment and people might think the information is credible because the person is famous, but if it is untrue, people generally have the ability to voice those opinions by commenting on the post. In this way, the public can keep communicators—professionals or average people—in check and can add their own thoughts to conversations about the environment.

"[Social media] has enhanced the ability to make us do better as journalists. If we did something poorly, we hear from our readers. There's a lot more feedback. It used to be the newspapers would land on your doorstep and that was the word in the media, and that's not true anymore. I've also seen really amazing climate communicators who have risen because of their own abilities, which is good to see too. It has given some scientist communicators big platforms and has brought them closer to the public," said Kate Sheppard, *HuffPost* senior enterprise editor and teaching associate professor at UNC–Chapel Hill's Hussman School of Journalism and Media.

Journalists are not the only people setting agendas. The ability of "average people" to communicate about the environment can be beneficial because people may trust those close to them more than people they don't know. Especially during environmental crises, like wildfires, I see my social media flooded with people I know reposting about them. Seeing people you know posting about the environment can influence your opinions and actions. Even though most of your friends are probably not professional scientists, if you see your friends talking about being nervous about the climate, you might be more likely to think you should also be nervous. People also might share their sustainable lifestyles with others, like by using metal straws and composting, which could encourage their friends to do the same.

The digital world is also different from traditional communications methods because it allows the audience and their engagement to be quantified to an extent, based on likes and comments. This measure contrasts with a general number of how many households bought a paper or viewed a newscast.

Sheppard said with the increase in social media, reaching an audience is a lot more competitive since people are not just reading the information a newspaper selects to send them. As a result, she said the questions become about understanding the audience and where they read journalism. She asks, "Is it a social audience? Is it a search audience? Is it people coming from my article? These are all different considerations than what we had before. What are the different platforms that we're engaging on? What is the audience on TikTok versus Twitter?"

TikTok is a rising platform for people to talk about the environment. Some climate communications among teenagers occurs on TikTok, including by kids involved in the Sunrise Movement. Some TikTok creators have garnered millions of likes by creating videos artistically comparing a clean world to a future decaying world.[49]

As times change, we must improve our strategies to effectively reach audiences.

"In the news, the inverted pyramid idea is to put the most important information first," continued Sheppard. I've learned in my classes that journalists should typically order information from most to least important in news stories. However, when we are competing against a lot of information and trying to reach new audiences, we may have to think about other ways to draw people into stories.

"There's more narrative and feature styles as well. We're not just competing against other news outlets; we're competing against all kinds of things on the internet, which is tough," said Sheppard. Her opinions and expertise highlight that communicators must find ways to draw in and provide value to audiences on social media. With it all at our fingertips, the abundance of information can make it more difficult to reach audiences.

Further, Maggie Kash, communications director of the Sierra Club, said social media has completely changed the landscape of communications and news media as we know it. She spoke as a communications professional, not representing the views of the Sierra Club, and said, "Environmental communications is part of a larger ecosystem of communications and media consumption" in the United States and worldwide.

---

49 Kris Bramwell, "TikTok videos spread climate change awareness," *BBC*, August 8, 2019.

The changes due to technology have been seismic. Kash said new technology has made the dissemination of information to the world and directly to members and supporters easier. However, she said new technologies also lead to difficulties since a vast amount of content exists online and people are inundated with information. She said communications now goes beyond the twenty-four-hour news cycle due to the amount of information available.

The Sierra Club's social media strategy depends on the platform because "not all platforms are created equal," continued Kash. She said various platforms "have different ways of operating and attracting different audiences" depending on their format. For example, different demographics are likely to engage with different social media sites. She said the Sierra Club pushes different campaign goals on different sites.

I also talked with John Bruno, marine biologist and UNC–Chapel Hill professor who communicates about climate change on platforms like an article in *The New York Times* and on a blog. Bruno said he questions what the right media for spreading information about climate issues is. He said *The New York Times* was probably not the right media for him to share the effects of climate change on coral reefs because it is mostly read by certain demographics and not likely to reach kids or climate skeptics. He constantly thinks about his platform, audience, and content.

Clearly, the explosion of social media provides a platform for people ranging from scientists to kids to talk about the environment, but effective environmental communications requires strategy.

MISINFORMATION AND INFORMATION OVERLOAD

A huge issue in recent news is pervasive misinformation. Social media can lead to false news, and studies show that most people are not good at determining true and false news.[50] Misinformation can be derived from and contribute to misconceptions about the environment. Sometimes people do not mean to spread false news, but information travels rapidly in the digital age.

Since media allows many people to share their voices today, deciphering what is true and what is false can be hard, possibly because of uneducated communicators or purposeful misinformation fueled by industries that will benefit from people misunderstanding environmental problems.

Climate communications expert Susan Joy Hassol explained social media "allows for the spread of misinformation and disinformation, for bots and trolls that undermine people's understanding and create doubt. We have to expand the possibilities for using those tools for good and try and find a way to decrease using them for evil. I would try to hold companies like Facebook and Twitter to account for allowing misinformation to be promoted on their channels."

Receiving only misinformation is not random. Hassol continued, "Everybody follows who they want to follow. And we can end up in our own echo chamber, where we're only listening to and amplifying the messages of those who we already agree with. So unfortunately, people who already are dismissive with regard to climate change are not exposing themselves to the

---

50   "'Fake News' Isn't Easy to Spot on Facebook, According to New Study," University of Texas at Austin, November 5, 2019.

voices on social media that could help correct that misinformation." She explained a way to overcome this challenge is to use "trusted messengers," like a person of faith talking about climate change with another person of faith.

Even when news is true, the perception is often still biased. With information overload, we experience selective bias because we pay attention to what fits our beliefs. For example, I'm interested in understanding climate change, so I follow a lot of environmental accounts. As an environmental studies major, I also have made a lot of friends through my environmental classes and clubs, so I am particularly exposed to environmental media when they share content. Therefore, much of the media I consume comes from people who believe in climate change and are very pro-environment. As a result, I might think the opinions I see a lot reflect how the general population thinks. Alternatively, somebody who isn't already interested in the environment might think these issues aren't as big of a deal or be unsure about how to talk about these issues because they think that nobody's talking about them since they don't see such content online.

Thinking about how people view information from their own lenses, I often wonder if the increase in media coverage of climate change protests has increased such perceptions and created more divide than good. Of course you should advocate for what you care about and feel is right, but we also need to be strategic in how we advocate for those causes to achieve our goals.

I voiced this concern to Gangi. He said, "In Europe, maybe it took a lot of students who were passive and made them less passive. So I would think even here, maybe in the US, it might have had that effect. So I think it's had a really positive effect among young people, taking people with positive values and making them a little more front and center. But I don't think it's really convincing to a lot of older people who kind of see it through a partisan divide."

The goal is not just to project environmental information into the media, but rather to frame it in such a way that more people can understand and support environmental activism to make it successful. Gangi said, "I think a big problem with some of the young environmentalists is there's a strong anarchist streak in parts of the progressive movement. [There is a] perception of environmentalists being anti-job, anti-capitalistic.

"When you're trying to grow a movement, you have to do two things. You have to get new people in your movement. But then you also have to convince others, and the power is still with the older generation. [You have to] reach out to older people who feel like environmentalism belongs to the other side and try to win them over," continued Gangi. I discuss more about effective protests in a later chapter.

Social media creates and perpetuates images and stereotypes of environmentalists that can make people turn away. The need to reach people who are unconcerned or in denial about the environment requires strategic (while truthful) framing to be more appealing.

One of the more unique and recent challenges with this explosion of digital communication is the massive overload of content we see every day. Of course, the internet has now been around for decades. But each year we witness newly emerging popular forms of communication.

Many people in my generation follow thousands of people across multiple platforms, whether on YouTube, Instagram, Twitter, Snapchat, or TikTok. The media we consume builds up the way we think. We don't necessarily realize how the content we're seeing is affecting us. Space for comments may help us discuss and reflect on information, but it could also lead to disagreements and increase divisiveness.

We need to utilize new, seemingly unconventional sources of information when communicating about time-sensitive issues like climate change to maximize our reach. Snapchat has exploded from just an app for sending pictures to a platform that gives news at people's fingertips while they're socializing. They learn about environmental issues even if they wouldn't regularly check a news website.

We also have creative ways of talking about climate and countless accounts we can follow to learn about issues. Some of them are more general, and some of them are specific, such as about waste or nature.

Another challenge with the online world is that it allows for continued greenwashing, which is a term environmentalist Jay Westerveld coined in 1986 that refers to brands marketing themselves as more eco-friendly than they are in reality.[51] Such inaccurate communication from seemingly trustworthy sources—businesses—can complicate consumers' knowledge of their own impact on the environment.

---

51   Adryan Corcione, "What is Greenwashing?" *Business News Daily,* January 17, 2020.

Over the past decade, climate scientists talking about the implications of their work when sharing their information with the public has become more acceptable. Communicators no longer always have to talk about the proof of climate change but can talk about what needs to be done and voice their informed opinions to the public through many avenues. Notably, the demographics of social media users are not the same demographics as the country, and we should gear strategies toward those who will see social media.

Ultimately, the digital world both complicates and enhances sending accurate environmental messages to the public, emphasizing the need for strategic communication. Social media is here to stay for the foreseeable future, and we should utilize it responsibly to amplify environmental messages. It can be beneficial as long as we follow social media posts with appropriate action. In the next section of the book, we will explore strategies to use in our digital and polarized state.

DISCUSSION QUESTIONS:

- Based on this chapter and your own experiences, do you think social media hurts or helps the environmental movement more?

- If you use social media, which sites do you use most? Does the content you see tend to reflect your own interests and opinions, or do you feel exposed to new ideas and viewpoints?

- How often do you think information you see online is false or unfairly framed? In what ways do you determine if a source is trustworthy?

PART TWO

PRINCIPLES FOR EFFECTIVE COMMUNICATION

4

REACHING AUDIENCES

—

Take a moment and think about your own beliefs about the environment. Do you believe in climate change? What makes you think what you think and perhaps take the action you already do against climate change? Can you recall any turning points in your belief or any conversations or media that affected you?

Effective communications comes down to individuals and what makes each individual believe and ultimately act.

If you're reading this book, you likely feel some level of concern about the environment. Communications consists of two sides: the communicator and the audience. We should consider how to best reach different audiences, whether communicating directly with an individual or talking to many people at once through mass communications.

I asked *Forbes* energy and environment journalist Jeff McMahon, "What is your intended audience when you talk about environmental issues; are you appealing to skeptical or disinterested audiences, or focusing on audiences already concerned about conservation? How do you respond to skepticism?"

He replied, "For each story I write, I imagine a specific audience or set of audiences and I write with them in mind. I generally don't write for skeptics, but I do realize that some readers will be skeptical. I know that some portion of my audience will care deeply about conservation and some other portion—since I write for *Forbes*—will care deeply about money. That concern about money can actually help get things done when economics align with environmental protection. So this awareness of audience helps me determine what elements to include in each story and in what order."

I then asked, "What would you consider to be the knowledge level of your audiences regarding technical environmental issues, like energy, and what is your strategy for explaining these issues while keeping articles concise?"

McMahon responded, "I assume my audience to be intelligent and to possess a general education but not to have particular expertise in energy or climate. So when I deploy a concept that a general audience will find difficult to understand, I have to consider how carefully and in what detail to explain it. A lot of detail can lose readers who either already know the information or don't care about the detail. In those cases, I can often link to a more comprehensive explanation."

As McMahon explained, effective communicators think about who they want to reach and how to best reach them.

Yale Climate Communications, part of the Yale Center for Environmental Communication, has established "Global Warming's Six Americas" that describe the segments of public opinion surrounding climate change.

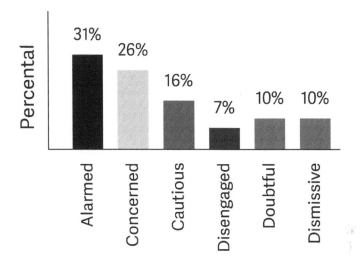

Global Warming's Six Americas.

These are the "Six Americas" (with percentages from 2019):

- Alarmed (31 percent): believe climate change is a significant problem and are taking action against it.

- Concerned (26 percent): believe climate change is a significant problem but think it is far off and have not taken much action against it themselves.

- Cautious (16 percent): are not fully convinced about climate either way and are unsure if humans are the cause of climate change if it is real.

- Disengaged (7 percent): do not know much about climate change.

- Doubtful (10 percent): believe climate change is not happening, but if it is, it is caused by nature and not humans.

- Dismissive (10 percent): believe climate change is a hoax and is not human-caused; strongly oppose climate action.[52]

A substantial number of people do not believe in human-caused climate change (bottom four categories), but many more people do view it as a serious problem. As you can see, more people are in the alarmed category than the bottom three categories combined.

HOW TO TALK WITH PEOPLE WHO ARE ALARMED OR CONCERNED

Maybe a lot of us think about those who do not believe in climate science when we think about communicating about the environment. But the reality is many people are already concerned about the environment, and we need to most effectively help them act upon their concerns.

Climate rhetoric expert Robert Cox recommended: "Years ago, we learned in communication studies that fear appeals work only up to a certain point and that strong fear appeals cause psychological barriers to be raised. It upsets people to the point that they have to put defenses against it, so they don't hear you at that point." We must find ways to motivate people to act, because telling the frightening facts without solutions makes the situation feel too hopeless for action to be worth it.

---

52 "Global Warming's Six Americas," Yale Program on Climate Change Communication, accessed July 7, 2020.

Communications comes in a variety of forms; you do not have to employ mass communications to be effective. Group or individual conversations can yield improved understanding and motivation to act against climate issues.

One individual leading such conversations and inspiring others to continue them is Don Holzworth, a professor at UNC–Chapel Hill's School of Public Health. When I interviewed him, he offered many strategies for effective environmental communications. We can use these strategies when communicating with people in the alarmed and concerned categories to increase their sense of self-efficacy and motivation to work toward change.

Holzworth describes himself as a serial entrepreneur. He has started many companies, including a global health consulting company that operated in sixty countries around the world. He is the entrepreneur in residence, and in this role he works with faculty and students who have ideas they think have commercial potential. He said he "had a successful career taking an idea from nothing to a large company."

While his career has focused on public health and entrepreneurship, Holzworth earned a degree in environmental science. He said he has always loved the environment and "all things related to the health of the planet," like clean air and clean water.

Holzworth and a co-founder run an event twice a year called Climate Conversations, an idea that came when he and a fellow serial entrepreneur—Lee Buck—were both concerned there was "not enough conversation about the threat of climate change to the human species." He said he saw almost no coverage of the link between the changing climate and the health and ability of humans to survive on earth. Holzworth said the planet will be fine, but the outlook for life is less certain.

As a result, Holzworth wanted to hold events to bring people together to converse about climate. He set up a nonpartisan community. He focuses on people who are influencers in the community, like business leaders, and has expanded the event to some students since they will ultimately solve these problems.

His purpose for holding these talks is to give people the opportunity to understand climate change, act on what they learned, and discover how to influence others. He said the theme of the events encompasses three words: learn, act, and influence. He wants to focus on those people who are "eager for knowledge and want to act."

Climate Conversations involves a presentation from a main speaker on topics like the science of climate change, communication around climate change, and energy transformation around the world. About 120 people attend each of these events, and they are by invitation only. Holzworth and his co-founder, Buck, pay for dinner out of their own pockets and make sure attendees do not perceive the events as fundraisers. He regrets that these conversations are not as diverse, since the community leaders attending are not as diverse. He said this discussion is not reaching the Hispanic population at all, and Black Americans are underrepresented.

Business leaders have taken action with their own companies to change the ways they handle waste as a result of these talks. They have learned the importance of electric vehicles and the transition off fossil fuels, and seen how alternative energy sources have created more jobs than other industries across the globe.

Holzworth said people find it enlightening to learn how to buy carbon credits to offset the carbon emissions caused by air travel. People can buy carbon offset credits through programs that fund projects limiting emissions, making up for the emissions from air travel.

Holzworth said another large change people can make individually is moving toward a plant-based diet. For example, eliminating beef from the planet would significantly reduce greenhouse gas emissions, not to mention the land and water usage that goes into meat production.[53] He is now very conscious of what he eats.

Additionally, reducing plastic consumption is part of mitigating our environmental impact. As he spoke, Holzworth held a bottle of orange-flavored Gatorade. He said he was embarrassed to have it, but he was on a liquid diet because of a medical procedure the next day. It was the first plastic bottle he had bought in a year because he and his wife are very conscious of not buying groceries wrapped in plastic.

Holzworth said the main ways to create an impact are to "communicate what you know and understand," to initiate

---

53   Richard Waite, Tim Searchinger, and Janet Ranganathan, "6 Pressing Questions About Beef and Climate Change, Answered," World Resources Institute, April 8, 2019.

conversations about environmental issues, and to "speak with your vote" by selecting leaders who are concerned about the problem and will take action.

He also recommended that young people talk to their parents to encourage climate awareness and action. Evidence suggests kids can influence their parents' view of climate change and affect their actions. A study in North Carolina found middle schoolers, particularly daughters, were able to influence parents, especially male and conservative parents, to be more concerned about climate change.[54]

Another tip is to take the time to write to congressional representatives and senators. Holzworth told me a local Citizens' Climate Lobby leader said a representative considers a handwritten letter as 1,000 votes. They are more valuable than phone calls or emails. Holzworth is a resident of Colorado, and he said one of his senators is not taking action on climate. He begins his letter with a thank you for a recent decision, explains his position, and responds if the staffer writes back without a satisfactory response.

While individual actions will not solve climate change, they can help people take steps to have more hope and motivation in attempts to rectify the environment. I discuss ways to overcome anxiety around the environment more in the following chapter.

---

54   D.F. Lawson, K.T. Stevenson, M.N. Peterson, et al, "Children can foster climate change concern among their parents," *Nat. Clim. Chang.* 9 (2019): 458–462.

HOW TO TALK WITH PEOPLE WHO ARE
CAUTIOUS OR DISENGAGED

Some people do not already understand climate science and the need to act urgently, but they are not against the idea.

We can see an example of communicating with this audience from when UNC-Chapel Hill environmental science and journalism student Brooke Bauman wrote for an internship with the Yale Climate Communications Office in Carrboro, North Carolina.

As part of her internship, she said, "I decided to write an article about a national park and the impact of climate change on national parks because I wanted to cater to the masses. The Great Smoky Mountain National Park I visited is the most visited national park in the whole United States—11 million a year—and it's free, which makes it accessible, and I wanted to engage people who have visited there." She wanted to help them "realize that climate change might impact how they experience it, so I looked for ways to get people excited about reading."

Bauman tried to think about different platforms and how to explain her mission for writing the article to draw people in if they weren't engaged. She wanted her article to be like an adventure story about national parks to capture their attention. While disengaged people do not represent the majority of Americans, she wanted to write something that would be interesting for many people.

Bauman received one piece of feedback that brings up an important point of discussion. One reader pointed out her article used a lot of "what if" scenarios and that the science was unclear about the impacts of climate change on mountain ecology. Bauman said, "I thought that was valid feedback," and that she was writing the article while a lot of science was ongoing. "I defended my decision to write it because I feel like people should be aware of the potential impacts of climate change even if we don't see them now so that we can try to take action now," she continued. We risk serious climate and economic consequences if we do not take action, and we must consider environmental issues even if the data is not completely proven yet.

Further, Bauman said, "The science is so complicated and it's so hard to explain the nuances. I think sometimes those can get lost in headlines when you're supposed to have a one-liner.... The combination of science and journalism is really interesting because you have to figure out how to communicate these nuances." As we think about audiences, we must provide details without overwhelming people—and we will discuss artistic and entertaining examples of environmental communications later.

When talking to people in the disengaged or cautious categories, the point is not just having to prove climate change but also explaining what can be done to combat it.

Bauman knows our society can take steps to mitigate climate change, like increasing our use of renewable energy. She continued, "I would encourage people to have this conversation with everyone even if they're not interested in it because it will get their wheels turning. Although awareness is not enough to save the world, it's the first step toward encouraging people to take action toward sustainability. Another thing to highlight about environmentalism is that everyone can incorporate environmentalism in some way. Even if you're in business, you can make business more sustainable. If you manage a construction company, you can get more sustainable construction materials; you can take public transit to your job."

Similar to when talking with alarmed and concerned groups, focusing on the solutions can appeal to more people; climate is not just an issue for them to be aware of but one that requires engagement. The need for action can make the issue relevant and encourage more people to listen.

HOW TO TALK WITH PEOPLE WHO ARE DOUBTFUL OR DISMISSIVE

Even discussing climate change with people in these categories does not need to turn into an argument. We have strategies to encourage doubtful and dismissive people to make sustainable choices despite factors like polarization and misinformation in the media that tend to cause challenges.

We should be empathetic with our audiences. With misinformation and information overload, I realize knowing what to believe can be difficult. We should not act like people who do not believe in climate change are ignorant. We also should remember that we do not have to encourage those we communicate with to be perfectly sustainable but rather to take steps toward sustainability. We should not attack nonbelievers nor should we tie climate change to other controversial issues. The goal here is not to turn everyone into a liberal politically but rather to make climate change something that people across the political spectrum understand and can come together to support.

One of the most effective strategies is to connect climate change to what people value, such as their health or children's future.

Robert Cox told me about a time in Kansas when connecting to values was successful for encouraging climate action. A nonprofit's goal was to encourage people to save energy. The nonprofit determined to not use the terms "global warming" or "climate change." Instead, the group focused on conservative values such as saving money from cutting power, being good stewards of God's Earth, and being patriotic by reducing foreign oil dependency. The campaign to save energy was not controversial since it was not climate-focused, but it did help to mitigate climate change.

Additionally, bringing local climate impacts into the conversation can help people understand the need for action.

"In Utah in Salt Lake City, there's actually a lot of support to address climate change because they've seen global warming has melted snowpack. And that's affected not only skiing or the tourism industry, but water sources for water intensive places like Salt Lake City and Ogden and so forth," continued Cox.

Don Holzworth, the public health professor who offered strategies for effective communications, said to help skeptics understand environmental issues, we should find common ground. For example, he will ask people if they appreciate clean air and water. Then he will ask if they understand that public health is threatened by the government relaxing pollution standards.

He said sometimes a person's faith can create challenges for their acceptance of climate change. For example, former WRAL meteorologist Greg Fishel is a Christian whose faith got in the way of accepting climate change. However, Fishel has come to accept the facts.

Fishel dismissed climate science for years until he realized his confirmation bias in 2005. He slowly began to consider that humans could be affecting the climate, started talking with other scientists about the subject, and began to voice the reality of climate change. He said he was called a "Marxist" and "Socialist" even though he had been a registered Republican for thirty years. Fishel now encourages others to learn and be willing to change their beliefs.[55]

Another tip Holzworth gave for conversing with skeptics is to ask them not if they believe in climate change, but if they accept the facts.

Holzworth said humans have an "innate ability to deny reality. As a serial entrepreneur, I've done things others told me I couldn't do because I denied reality." He said this ability to deny reality has led to advances in science and technology, but in the case of understanding climate change, this ability "may lead to our demise as a species."

---

55 Greg Fishel, "How a North Carolina meteorologist abandoned his climate change skepticism," *Columbia Journalism Review*, November 13, 2017.

But he has been successful encouraging people to take action for the climate. For example, he said Stacy MacIntyre, CEO of a hospital system, made many changes in the way supplies are ordered and disposed of in their hospitals after attending his event. She wants to install electric car stations in the hospital parking lots, has decided her next car purchase will be an electric car, and has talked to her key executives about what she learned from conversations. Thus, his event has led to people "thinking and acting differently."

Climate conversations also taught a representative from the Triangle Community Foundation that she could advise donors to use philanthropic dollars to give to organizations that support sustainability; this is yet another way conversation can encourage a person to integrate sustainability in their life and organization.

The solution comes down to finding how you can agree rather than disagree with people to encourage them to support sustainability. Holzworth said he found common ground when discussing environmental issues with his brother. His brother is a climate skeptic but a "fanatic about recycling" and will walk two miles to recycle something. They can connect based on their shared value of protecting the environment.

Holzworth said sticking with indisputable facts is important. Don't talk about saving the planet, because it will be fine—instead talk about potential harms toward humans' way of life. An example would be when the air in Delhi, India, was so bad that schools had to close and thousands of people had to wear gas masks.

Since I talked to Holzworth, people in the United States have indeed had to close schools and wear masks—and many people understandably do not like this way of life. The COVID-19 pandemic highlights that global events can cause death, chaos, and the need to adapt to new ways of life. Taking actions like mitigating climate change and pollution now is better than tackling the challenge of adapting to it later.

Personal relationships are critical for convincing people to accept climate change.

"A lot of what research has shown is that the messenger is really important. There are certain messengers that people really trust. People trust their weathermen; people trust their doctor," said Kate Sheppard, *HuffPost* senior enterprise editor and UNC–Chapel Hill's Hussman School of Journalism and Media teaching associate professor. This point highlights the significance of Fishel's change of mindset and advocacy for learning about climate change.

"As a journalist, obviously the science is the utmost importance and getting that into the research and their stories into the work is very important, but also getting regular people that your reader can identify with in there as well. And always connecting big picture science or policy stories with actual impacts on people," she continued.

Another strategy Sheppard espouses is connecting climate to people's values. For example, "if your reader cares about the military, then let's talk about sea level rise affecting bases, or military readiness, or where we have to intervene in foreign countries. If you're someone who cares about inequality, or poverty, [you could discuss] the health effects on people who are [experiencing] those issues. Most directly, if you're worried about kids, [then connect climate] with kids. Basically, try to find the values people already have and how you tap into that with the issues and awareness that you're trying to bring about environmental or climate issues." Additionally, communicators can connect to religion; for example, Christians often care about reducing inequality and poverty. Understanding that climate affects some people more than others can make them change their minds about being skeptical.

She gave me a personal example. "I grew up on a farm, and my dad is an economic conservative, he is Republican, but he is a farmer and a lot of climate stuff connects with him because he sees it in his daily life. And I also heard a number of farmers who maybe [you] wouldn't expect to be thinking about climate change, but they're thinking about it, because it connects with their work."

Yet scientists also ought to be able to communicate about their own work. Unfortunately, the public can fear or distrust scientists if people do not understand who scientists are and how they work. One particular issue is the meaning of uncertainty. Scientists may say an outcome is uncertain if they do not have proof of the exact outcome yet, even if some outcomes are highly likely. Uncertainty should not prevent people from sharing information. On the other hand, the public may think uncertain outcomes are not worth considering. An example of confusion around uncertainty is the impact of climate change on hurricanes. In recent years, hurricanes have become larger and slower-moving. Scientists can determine climate change is affecting such weather patterns but cannot tell exactly how climate change affects particular storms. As a result, conveying weather risks of climate change to the public can be a challenge.

"Our knowledge is constantly evolving, so something that was uncertain or unclear ten years ago is now much more clear," said Sheppard, showing that uncertain information can indeed be true and relevant to share. While scientists may accept ideas, like gravity, as true, the public may be more apprehensive about something they hear is uncertain. Sheppard told me scientists can attend trainings to help them most effectively communicate with the public when science is uncertain.

Sheppard discussed ways people can communicate about the environment themselves with others around them. She said an effective communications strategy requires "deep listening and trying to understand where people are coming from" so that communicators can best understand how to connect to people and make their points resonate.

Connecting to values can draw in people who are doubtful and dismissive about climate change. However, the audience at the mass communications level is unlikely to be split into just one or two categories. We still have to think about appealing to all groups of people to some extent.

"It is always a tension in news between what our current audience wants versus how might we get a different audience," shared Sheppard. "Personally, in my career I went from working at an environment-focused site where everybody who read that site cared about the environment to being the environment reporter out of a politics-focused magazine, to the main environment person at a bigger site, and now I edit the environment and econ team so I have to think a little bit more broadly, not just the environment. I'm trying to reach a very big audience. Every time that the job has changed, it's a little bit of a different calculation about who is my audience, who am I reaching, who should I be trying to reach and who is just not going to be [engaged] no matter what I do."

OTHER STRATEGIES AND CONSIDERATIONS

Storytelling is a major strategy for helping people resonate with environmental issues. In April 2019, Stephen Friedman, former president of MTV and Emmy and Peabody Award-winning creator of social impact campaigns, visited the School of Media and Journalism at UNC-Chapel Hill to discuss the media as a superpower for the public good. He said humans are irrational and that journalists can "rarely move minds by leaning into facts," but he espoused storytelling as an effective strategy for communicating about social issues. Finding a common value is helpful, followed by the use of storytelling devices, like anecdotes, around that value. Stories define us, but emotion is necessary to connect to the audience. Although fear can make people frozen rather than spurred to action, Friedman said that "the best journalists translate fear into something equally visceral." Also, he said journalists cannot "rest on morals that make sense" when trying to reach an audience and inspire change. Holzworth continued this idea when he said that communicators must maintain truth and authenticity by telling stories in a balanced way to gain trust, especially since most people do not identify with activists.

A common question: how do we talk to kids about climate change? We must keep it part of the dialogue, including the fact that marginalized communities disproportionately face environmental problems—some kids have experienced this but others haven't. We can start by instilling a care for the environment by bringing kids outside from a young age. An NPR podcast recommends encouraging kids to love nature, like bugs, before making them scared about climate change. It also recommends that we "focus on feelings" when discussing environmental problems, with the understanding that kids care about nature, like sea turtles dying from plastic. We can help kids deal with feelings of being scared by listening and motivating them to take action, like attending protests or picking up trash.[56] We should keep in mind that many young people are already anxious about the environment, so we generally can follow the advice for talking with people who are alarmed and concerned. Specifically, that means providing them those action steps to keep them hopeful. Then they can continue talking about climate change and fighting for a better future.

Another consideration and potential challenge when reaching some people, particularly men, is cultural norms that can make caring about the environment seem feminine. "Toxic masculinity" is hurting the environment.[57] For example, an experiment published in *Scientific American* showed both men and women perceived reusable bags as more feminine and single-use bags as more masculine. A solution is to reaffirm

56   Anya Kamenetz, "How to Talk to Kids About Climate Change," *NPR*, October 24, 2019.

57   Joe McCarthy, "Toxic Masculinity Is Killing the Planet, Study Finds," *Global Citizen*, May 23, 2018.

someone's masculinity when discussing sustainable choices. The article says, "In one experiment, men who received feedback affirming their masculinity were more interested in purchasing an eco-friendly version of a cleaning product. Men who feel secure in their manhood are more comfortable going green." Additionally, this article recommended, "green products and organizations can be marketed as more 'Men'-vironmentally-friendly, with more masculine fonts, colors, words, and images used in the branding."[58] Not all men think in these ways, but people should consider potential effects of gender stereotypes and pressures on receiving and acting on messages about the environment.

I think back to examples in my own life of how communication can be engaging and impactful. I saw my local news station, WRAL, post a video on its website titled "Science experiment: how carbon dioxide affects oceans." This video appealed to me because I figured it related to the looming issue of climate change, since carbon dioxide is heavily responsible for it, but the title of the video itself did not spark fear or political tension. Although tension can make stories interesting, looking at a more informative rather than persuasive communication encouraged me to pay attention. The video was just over three minutes, which I thought was the perfect length to help the audience understand the explanation of the science behind climate change impacts on the ocean, while being brief enough to gain an audience.[59]

58 Aaron Brough and James E.B. Wilkie, "Men Resist Green Behavior as Unmanly," *Scientific American*, December 26, 2017.

59 Mikaya Thurmond, "Science experiment: How carbon dioxide affects oceans," *WRAL*, January 4, 2020.

The "Six Americas" are broad categorizations, and of course everyone thinks differently. Nevertheless, we can understand our audiences in order to best appeal to them, increase their understanding, and encourage action based on how knowledgeable and concerned they already are about the environment.

DISCUSSION QUESTIONS:

- Which category of the "Six Americas" do you think describes you? Which category do you think describes most people you know?

- Have you ever tried to convince someone that climate change is real and/or that we need to take climate action? How did it go? If it did not go well or you have not tried before, what could you do to persuade people about the importance of climate action in the future? If it went well, why do you think you were successful?

- What are some of your values that connect to the need for climate action?

5

PSYCHOLOGY AND TYPOLOGY

———

My heart felt like a bowling ball sliding down my chest. I was scanning a web page in my climate change and psychology class. The page showed that at the current rate of global emissions, we will experience severe levels of species loss, food loss, drought, wildfires, sea level rise, and more. I had been advocating for the environment for years, but sometimes it really hits me that undoubtedly we are facing and will continue to face terrible effects of climate change on both us and the environment. I wondered if there was any point in doing environmental work.

Around that time, I went to a panel called Solastalgia, which is a term for the stress resulting from environmental change. "Solastalgia" is a term coined by philosopher Glenn Albrecht.[60] Jaya Kuic, a geography student at UNC–Chapel Hill, planned

———

60 Georgina Kenyon, "Have you ever felt 'solastalgia'?" *BBC*, November 1, 2015.

the panel. At this panel, students across multiple disciplines at my school discussed their experience and research with mental health issues and the environment.

Being so entrenched in the "environmental world," taking environmental classes, and spending a lot of time with environmental organizations can be taxing. I repeat the word "environment" many times a day, especially when people ask me what classes I am taking or what I spend my time doing. I have to admit that the environmental issues I spend time learning about and attempting to mitigate do get me down sometimes. Attending this panel showed me that other people feel the same way.

I learned that environmental scientists and students studying environmental science are at risk of developing mental health problems from focusing so much on challenges and issues that impact the world. This fact raises the importance of considering mental health when communicating about the environment. Describing to people the significant environmental problems when solutions are somewhat out of their personal reach may cause harm to them. We should not blame individuals for large-scale issues. Also, we need to point out resources for people who do face eco-anxiety or grief.

The panelists discussed the importance of giving people tools to manage their feelings. For example, people can make sustainable choices in their own lives that will not save the world but will help them do something, and they can spend time with others in environmental groups to feel empowered.

A lot of climate change communications is focused on convincing skeptics of the science, but we also need to understand that many people are highly concerned about the environment, especially young people. Most American teens are frightened by climate change, a 2019 poll finds, and about one in four are taking action. The poll shows that 86 percent of teens think human activity is causing the climate to change and 57 percent of teens feel afraid of climate change. Also, 54 percent of teens said they feel motivated about climate change, and 43 percent said they feel helpless (they could choose multiple options among a list). The poll sampled teens across the United States.[61]

Martha Crawford, a therapist in New York City, documented some of this anxiety surrounding climate change through her "Climate Dreams Project." She collects and posts dreams people have related to climate, like natural disasters.[62] Climate is seeping into dreams, conveying the seriousness of mental health effects of climate.

Further, 66 percent of people surveyed in a 2019 poll (not just teens this time) were at least "somewhat worried" about global warming, according to the Yale Program on Climate Change Communication.[63]

Many people are concerned about the impact on people other than themselves. "Almost all of the young people interviewed for

61   Liz Hamel, Lunna Lopes, Cailey Muñana, and Mollyann Brodie, "The Kaiser Family Foundation/Washington Post Climate Change Survey," Kaiser Family Foundation, November 27, 2019.

62   "The Climate Dreams Project," accessed July 7, 2020.

63   Yale Program on Climate Change Communication and George Mason University Center for Climate Change Communication, "Climate Change in the American Mind," November 2019.

this article said they were struggling with the ethical implications of having children," said writer Avichai Scher in the NBC article "'Climate grief': The growing emotional toll of climate change."[64]

I covered a climate strike on UNC-Chapel Hill's campus for my journalism class in September 2019, and I spoke to some high school students in attendance who told me they were worried about having kids in a world with an uncertain future due to climate change. This issue is scary for many people in my generation, as we are growing up in a world where we do not know how long the world will be able to maintain human life, or at least the quality of life we have now in many places.

Despite increasing fears about climate, we will examine strategies to communicate about the climate crisis with different groups of people. We need to understand how to reach different audiences, including young people and both those who want to do something about it and those who feel the situation is too hopeless to change it.

From considering eco-anxiety to thinking about public response to most effectively reaching audiences with diverse personality types, we must use psychology strategies when communicating about the environment. The nexus of the environment, communications, and psychology provides further insight for reaching audiences.

Like climate rhetoric expert Robert Cox mentioned, we cannot just tell people that climate change is already happening and

---

64   Avichai Scher, "'Climate grief': The growing emotional toll of climate change," *NBC News*, December 24, 2018.

that it is too late to stop it if we want them to take any action toward mitigating it.

We must maintain and spread hope. Community is important. I witnessed the importance of community when I wrote for Heart of Waraba, a nonprofit that connects sustainable entrepreneurs around the world to build a sense of community and provide hope for working to solve the climate crisis through innovation. Working on solutions to environmental issues can be lonely, so we should look after our own and each other's mental health.

Beyond recognizing eco-anxiety, we must think about rhetoric and the best ways to encourage both change in opinions and action.

Language is important when discussing climate change... or the climate crisis... or the climate emergency. Rhetoric matters. Public opinion analyst Frank Luntz said certain words better lead to action. For example, he suggested replacing "threat" with "consequences," green energy "jobs" with "careers," and to not use the crisis loosely when discussing global issues. Greta Thunberg has drawn attention by using the phrases "climate breakdown" and "climate emergency." The climate problem is not just scientific, but also linguistic.[65]

Additionally, Luntz suggests using the word "climate change" rather than "global warming" to encourage nonpartisan climate action because the former is more encompassing of the

---

65   Kate Yoder, "Frank Luntz, the GOP's message master, calls for climate action," *Grist*, July 25, 2019.

effects of greenhouse gases and more difficult to dispute.[66] Global warming can feel scarier, and people tend to turn away from challenges that feel too daunting.

People can easily become numb to or dismissive of climate issues when they hear about the likelihood of "climate change" all the time without understanding the true details or urgency. Terms like "climate change" develop connotations, pointed out Anthony Leiserowitz, director of the Yale Program on Climate Change Communication. Some of these connotations are "elitist," "liberal," "socialist," and "hoax."[67] Therefore, we need to focus less on facts about the problem and more on providing viable solutions.

We can overcome fear by following up discussions of problems with explanations that solutions exist that could lead to better outcomes than utter catastrophe. We can still mitigate climate change by taking steps now, such as increasing energy efficiency and more quickly transitioning to clean energy.

Another challenge is scientists must be truthful and say when potential climate consequences are likely but not certain, yet the public does not respond as seriously to consequences that are uncertain. The second volume of the fourth National Climate Assessment uses the word "likely" 867 times, conveying that negative outcomes on the environment are uncertain yet realistic. The assessment includes the need for scientists to emphasize that extreme changes in climate are going to

---

66 Ibid.

67 Dan Zak, "How should we talk about what's happening to our planet?" *The Washington Post*, August 27, 2019.

happen rather than convolute information about projections with details on uncertain specifics in order to better communicate about environmental issues.[68]

One model that conveys how people think about climate change is the climate change threat index, developed by researchers Jason T. Carmichael and Robert J. Brulle. This index shows that major media is the greatest factor for how much the public views climate change as a threat, among other factors like the weather and economy. Contrary to what scholars previously thought, the researchers found that extreme weather events do not significantly shape national public opinion on climate change.

According to Carmichael and Brulle's 2017 research, three other factors affect the level of public concern about climate change the most: (1) the quantity of media coverage on climate change; (2) the competition climate change faces to be on the media agenda; and most importantly, (3) the amount of attention Congress gives to climate change and controversy around that.[69] As we continue to think about best ways to reach audiences, especially with the goal of influencing public understanding of and desire to take action on climate issues, we should recognize that media coverage plays a huge role.

I learned about other applicable psychology concepts in my climate change and psychology class. The theory of reasoned action shows people have beliefs about an idea that lead to

---

68 Ibid.

69 Jason T. Carmichael and Robert J. Brulle, "Elite cues, media coverage, and public concern: an integrated path analysis of public opinion on climate change, 2001–2013," *Environmental Politics*, December 5, 2016.

an attitude about it, which leads to intention to do anything about it, which leads to behavior. Additionally, the theory of planned behavior shows that a person's attitude and subjective norm (what they think others feel about an idea), along with perceived behavior control, lead to intention and behavior. Combining these theories conveys that both an individual's attitude and perceived control influence how they behave. Thus, we need to make people not feel hopeless about the environment if we want them to make sustainable choices.

Although public opinion needs to shift to encourage policy changes, ultimately we must encourage people to change their behavior.

In a TED Talk, behavioral neuroscientist Tali Sharot explained three principles for encouraging behavior change. The first is social incentives. She said people care a lot about others' opinions and how they compare to other people. She highlighted a study in the UK in which people were more likely to pay their taxes on time if they were told nine out of ten people did so. Therefore, rather than discussing the environment as a disaster most people are not doing anything about, talking about what others are doing could encourage individuals to feel that action is within reach and they should join in.

The second principle is immediate rewards. Sharot said people associate a behavior with a reward until it becomes a habit. Thus, showing people that behavior changes, such as using reusable items, are not inconvenient but rather associated with a reward—like lower costs and less time spent shopping—is a good strategy.

The third principle is progress monitoring, when you "highlight the progress, not the decline." Sharot stressed that warnings have a limited impact and often lead to freezing or fleeing rather than fighting. She said people tend to register positive information rather than negative information, creating a positive self-image. This concept can apply to the environment when people choose to disregard news about environmental crises and not feel the need to change their own habits.[70]

We should consider psychology when thinking about an audience, large or small. Evidently, we can use psychology in a variety of ways to better communicate about the environment.

While these are some more established psychological ideas for discussing climate change, we also have some ways to use personality typology to reach audiences.

I have a fervent interest in the Myers-Briggs personality types, which characterize individuals as one of sixteen types based on four spectrums. When I first learned about the types, I was stunned to discover how accurately I felt the INTJ (introverted, intuitive, thinking, judging) type described me in some ways. As much as I relish better understanding my strengths and challenges through learning about my personality type, I also enjoy discussing personality types with other people or helping them find their personality type. I have a happy memory of my friend's wide, gleaming eyes over her excitement at gaining such insight about herself after I gave her a brief test to find her type.

---

70   Tali Sharot, "How to motivate yourself to change your behavior," YouTube, October 28, 2014.

At UNC–Chapel Hill, I participated in an event called SPLASH, where college students teach high school students about various unique topics for a day. I taught a class about using the Myers-Briggs types for introspection. I also gave a lecture to a small group of students at my high school about these types. When I talk to students about these types, they tend to tell me the types accurately reflect them and give them helpful insight on their outlook and how it differs from others. Students have told me they can better understand and empathize with members of their family or friends by recognizing that people think in different ways and have different needs, whether it be alone time or understanding ideas in logical ways.

Of course, the Myers-Briggs tests do not prove anything about an individual. The types are more of a categorization than an exact science.[71] I don't fit every characteristic of my type, and we should not let our types define us. For example, INTJs are known for being behind-the-scenes and quiet, yet that does not stop me from advocating for environmental action. However, I find it helpful to read about my type and gain introspection about my inherent traits. My experience shows me they can be helpful for people to understand themselves and each other.

People with different personalities may engage differently with climate issues. Not everyone knows their type, and on a mass communications level, you are speaking to a variety of types. But understanding an individual's personality type is a way to build on connecting to their values and being best received when talking about the environment.

---

71 N'dea Yancey-Bragg, "Here's why people still take the Myers-Briggs test—even though it might not mean anything," *USA Today*, May 6, 2019.

The middle two letters of the four-letter types are especially impactful for communications because they describe the ways people process information about the world around them.[72] If someone is an S, meaning sensing, they think in more practical ways about the reality of a situation. On the other hand, a person with an N type, meaning intuitive, is more imaginative about possibilities. The third letter is either F for feeling or T for thinking. This letter represents the ways people make decisions. A person with an F as their third letter may make decisions more on values and could be more empathetic, compassionate, and focused on keeping harmony in a situation. A person with a T might focus on scientific facts and logic, like weighing the pros and cons of a situation.[73] (The first letters are I for introverted and E for extroverted; the last letters are J for judging, meaning people prefer organized plans, and P for perceiving, meaning a person is more spontaneous, to briefly describe these types.) People can be in between each of the two letters.

When talking to an SF person, appealing to pathos can be helpful because this type relies on emotion to make decisions. Since they are also practical, showing the immediate benefits or consequences, tied in with emotional stories, is ideal.[74]

The ST type is very straightforward and objective, necessitating clear and logical reasoning, as well as specific benefits and consequences, when talking to someone of this type.

72 "Most Effective Communication Strategies with Various Personalities," HRPersonality, accessed July 11, 2020.

73 "Personality Type Explained," Humanmetrics, accessed July 11, 2020.

74 "Most Effective Communication Strategies with Various Personalities," HRPersonality, accessed July 11, 2020.

Visuals and charts can help display the situation to effectively convey it.[75]

When talking to an NF person, laying out the facts and appealing to their intuition to help them understand why you think something is a problem or why a certain course of action is best can be effective, as can talking about concepts and theories, since people with "N" as their second letter think abstractly.[76]

When talking to an NT person, you should think about concepts and theories and also include many facts and how an issue fits into the bigger picture.[77] As an INTJ, I agree that understanding the purpose of communications about a specific issue and how it can be important on a broader scale is helpful. For example, if reading about ocean plastics, I want to know the facts on how much ocean plastic exists and how it is hurting ecosystems at a broader level, as well as the possible causes and solutions.

A person with the SF type, on the other hand, might respond well to a story about animals found dead and filled with plastic and then a number of how many animals have been found like this. It's not that an NT person doesn't have a heart, but certain ideas resonate better with different people. I still feel sad when I read about plastic harming animals, and I do want to do something about it.

---

75 Ibid.

76 Ibid.

77 Ibid.

People communicating about climate change are not representative of the general population in terms of personality types. A research study found that, although the proportion of introverts to extroverts did not vary greatly from climate researchers to the general population (54 percent of researchers were extroverts and 46 percent introverts compared to 49 percent extroverts and 51 percent introverts in the National Representative Sample), the other three parts of the Myers-Briggs types did. Of climate researchers, 82 percent were intuitive and 18 percent were sensing, compared to 73 percent intuitive and 27 percent sensing in the National Representative Sample. Additionally, 49 percent were thinking and 51 percent were feeling, compared to 40 percent thinking and 60 percent feeling in the National Representative Sample. Most significantly, 76 percent of climate researchers were judging and 24 percent were perceiving, compared to 54 percent judging and 46 percent perceiving in the National Representative Sample.[78]

This research study explained one implication of these differences is that the greater amount of people who are the N type means these people might be biased to discuss future climate impacts. People of the S type, on the other hand, prefer to know what is happening currently and near them, rather than possibilities. Also, since more F types exist in the general population, T scientists should remember to include values and emotion when discussing science (while not skewing facts, of course). Further, the J type is more decisive, so the

---

78   Susan Weiler, Jason Keller, and Christina Olex, "Personality type differences between PhD climate researchers and the general public: implications for effective communication," *Climatic Change* (August 27, 2009): 237-241.

larger proportion of J scientists shows many of them may want to come to a conclusion rather than leave room for uncertainty, whereas the general population may focus on room for doubt.[79] Although communicators should be clear that uncertainty is different from unpredictability, they should explain the details to show that multiple possibilities exist.

Whether speaking to individuals or operating on a mass communications level, we need to understand that the personality types of our potential audience may vary from our own.

Another way to more effectively communicate with someone based on their personality is using enneagrams. There are nine enneagram types describing someone's personality. Each type copes differently with emotions, including anger, doubt, or fear. These types include (in order): the Reformer, the Helper, the Achiever, the Individualist, the Investigator, the Loyalist, the Enthusiast, the Challenger, and the Peacemaker. Shortly after I first read about the nine enneagram types (I am type one), they exploded across social media. I know many people who feel they have an enneagram type that applies to them.[80]

Just like the Myers-Briggs types, they are not an exact science. We should not put people in boxes or overgeneralize, but knowing a person's enneagram can help us understand their communication style, as well as their values and way of thinking, in order to better reach them when spreading

---

79 Ibid.

80 "How the Enneagram System Works," The Enneagram Institute, July 11, 2020.

environmental information. For example, someone who is type one—the Reformer, also known as the Perfectionist—may be detail-oriented and focused on right and wrong. Because they worry about being perfect, they may focus too much on jumping to a conclusion about what is "right" or what "should" be done, rather than considering alternatives.[81] This observation reminds me to focus on sharing facts with people and allowing them to come to their own conclusions about what to do about the environment, rather than just telling them what they "should" do.

Type nine—the Peacemaker—worries about maintaining harmony, so focusing on not offending people can be good, whether by not using personal identifiers like "you" or even by using all lowercase letters.[82] While calling people to action is important, not upsetting certain types, like type nine, may be better in order to reach them. Obviously, communications often targets multiple people, and we do not normally know our audience members' enneagram types. But, when communicating with a purpose, we can be cognizant that communicators have certain tendencies and also that people in the audience think in varying ways.

That advice doesn't just go for personality types. We can't read minds, but if we want to effectively communicate, we can understand how different groups feel about the environment and perceive information and base our strategies off that.

---

81   Lily Yuan, "The Enneagram Type One—The Perfectionist," *Psychology Junkie*, September 25, 2019.

82   Lily Yuan, "Here's How You Communicate, Based on Your Enneagram Type," *Psychology Junkie*, September 10, 2019.

DISCUSSION QUESTIONS:

- How do you feel when you think about climate change? Share any instances when you felt more or less motivated to take climate action.

- What would you say to people facing eco-anxiety to help them overcome this feeling and start to take action or continue taking action to help the environment?

- Do you know your Myers-Briggs personality type and enneagram? If not, many online tests exist. How might these categories influence how you perceive information about the environment?

6

IT'S NOT JUST CLIMATE CHANGE

———

Protesters sat in the streets in Warren County, North Carolina, trying to block dump trucks from delivering a toxic chemical, polychlorinated biphenyl (PCB), to a landfill. The year was 1982. The state government had placed a landfill in a mostly Black community, despite residents' concerns about the waste entering their water. The protesters were unsuccessful, and the toxic waste did end up in the landfill. However, the protest gained media attention.[83]

This event marks what many people consider to be the beginning of the environmental justice movement, though environmental injustices, and some protests, had been happening long before. The movement started not too far from where I live in North Carolina. Sadly, environmental challenges often affect certain communities more than others.

———

83   Renee Skelton and Vernice Miller, "The Environmental Justice Movement," NRDC, March 17, 2016.

Environmental justice occurs when "all people can realize their highest potential, without interruption by environmental racism or inequity." However, environmental injustice occurs when some people experience more of the burdens of the state of the environment than others. Similarly, environmental racism occurs with a "disproportionate impact of environmental hazards on people of color."[84] Warren County receiving the toxic chemicals is an example of environmental injustice and environmental racism, since an already marginalized group experienced worse outcomes compared to other groups of people.

Environmental and climate justice issues are numerous. Many people think of the environment as a generational problem because later generations face the effects of how their ancestors treated the environment. Another issue is that those responsible for environmental impacts often do not face the same extent of consequences. Wealthy people are responsible for greater consumption and greenhouse gas emissions, yet they can more easily adapt to a changing climate. Disadvantaged groups tend to consume less and contribute to climate change less, but they may have less money and ability to adapt to the environment. For example, they may have no choice but to live near less desirable places, like landfills and factories. They also may not be able to afford rising food prices or move from places that flood more often due to climate change. Notably, sustainable products ranging from ethically produced clothing to electric cars, which would help prevent the environmental problems they face, may be less accessible for these groups.

---

84 "Environmental Justice and Environmental Racism," Greenaction for Health and Environmental Justice, accessed July 5, 2020.

The attention given to racial injustice in 2020 following the deaths of Black Americans including Ahmaud Arbery, Breonna Taylor, and George Floyd, among many others, has further brought to light the connection between racial justice and climate justice.

Climate activist Elizabeth Yeampierre, co-chair of the Climate Justice Alliance, explained that climate justice is impossible without racial justice. For centuries, white Americans have exploited Black Americans the way people exploit resources at the cost of a healthy and sustainable environment. Our economy, focused on consumption, has been built on exploiting workers and the land. One of her examples of a problem with neglecting environmental justice in the environmental movement is that building windmills can reduce carbon emissions, but the materials are often carried on ships that use diesel and release pollutants into communities. Those communities face negative environmental consequences as the result of environmental work, highlighting the broad perspective needed when working and advocating to help the environment. Further, Yeampierre noted the sad point that the communities that suffer from respiratory diseases due to poor environmental quality are the same communities where people cannot breathe due to police brutality. Of course, other demographic groups face oppression and environmental injustices as well. To truly fight for climate justice, we must fight for all people.[85]

Environmental communications often focuses on natural resources and not the people affected, especially when certain

---

85 Beth Gardiner, "Unequal Impact: The Deep Links Between Racism and Climate Change," *YaleEnvironment360*, June 9, 2020.

people are affected most by generations of exploiting the environment. Those of us advocating for environmental action must remember to include all communities.

I talked to Diamond Holloman, a PhD candidate at UNC–Chapel Hill, about the best ways to communicate about environmental justice issues. We discussed the necessity to be truthful and direct, as well as framing when sharing stories about people facing environmental justice issues.

"I got interested in this work when I myself was an undergrad," said Holloman. "I lived in New York. I was going to New York University when Hurricane Sandy hit. And at the time I was an RA, or a resident assistant, at one of the dorms, kind of the higher and fancier dorms. And my family is from New York. We live in Bedford Stuyvesant in Brooklyn and I knew other people in Coney Island. And so the hurricane hits. And I remember being concerned about my family members not being able to get out to the stores, the power was off for days, the food in our fridge was really bad, and all these different issues with flooding."

She said the situation made her think of Maslow's Hierarchy of Needs, which depicts how people must fulfill their basic needs before they can focus on their psychological needs and ultimately their self-fulfillment needs.[86] "So I'm thinking about really those base [needs] like food, shelter, these kind of lower-on-the-pyramid needs." She noticed that a lot of students, who were mostly of a higher socioeconomic

---

86 Saul McLeod, "Maslow's Hierarchy of Needs," SimplyPsychology, March 20, 2020.

status and could afford to live in the nice dorms, had concerns like washing their hair and getting their nails done when the power went out, which were concerns higher up on the pyramid.

"I remember thinking how crazy it is that we're occupying the same space, but the concerns I have for my family and friends in Brooklyn are so different from the concerns other people have for themselves and their families," reflected Holloman.

Holloman said this difference reflects that different outcomes follow disasters. It reminds her of a time a few years before that when a blizzard came and her family was snowed in for days, whereas wealthier Manhattan had plowed the streets immediately. She said that these events, which caused different effects for different people, got her thinking about this topic and influenced her grad school work.

She conducts "ethnographic, qualitative work in environmental justice and looks at the ways in which different groups experience disasters, specifically hurricanes." Since 2017, she has been talking to people in eastern North Carolina and learning about their experiences. She looks at the ways municipal governments handle disasters like Hurricane Matthew in 2016 and Hurricane Florence in 2018, and how solutions are provided—or not.

People talk about inequities across society often, but they are often in journalists' words. Holloman thinks that "there is a lack of these communities being able to talk about themselves and express themselves and their experiences in ways that are authentic and legitimate for them. I think very often it's quite easy for someone outside of the community to come in, talk to people, and then just report back to all these different news outlets. And in their stories, they're writing about these people as subjects, as opposed to letting new people talk about their own experiences in their own words, in their own framings, which is really important."

"I walk a very thin line when I go in because I am an outsider, and I'm literally writing a dissertation, in which I have to analyze other people's words," continued Holloman. "I'm doing everything in my power to put their stories at the forefront and *in their words*, and in their framings." She shares her writing with interviewees to ensure she accurately recounted their views. "It's been very rewarding, and I'm glad that I get to do this process. I'm glad I get the space to do something that's so community-driven and collaborative."

Holloman raised the importance of letting people speak for themselves and using their direct wording and quotes as much as possible. In addition, following up with people to ensure accurate sharing and framing of their stories is evidently crucial. At NYU, her undergraduate school, she double-majored in environmental studies and journalism. Holloman knows sharing stories ahead of time is not standard journalistic practice, but she said doing so is important when integrating more qualitative research values.

Journalists and researchers can fight injustice by writing stories that showcase experiences of people facing injustice and by sharing those stories with policymakers and politicians who could directly help those people. Holloman said she is "trying to uplift the spaces that are already inhabited, the voices that are already there, the experiences and the stories that are already there, trying to put them in spaces that they aren't typically." Readers not facing injustice can help by sharing the stories and voices of people facing injustice as well. That can create change, which echoes the idea that effective communications can lead to tangible progress.

Further, Holloman said she thinks a misconception exists that people facing environmental issues don't fully understand them. Based on her research, she explained, "intuitively, they understand how politics and ecology work together in these spaces and how they intermingle and intertwine into their lived experiences. This is how community members will talk about their lives, so they fully understand what they're going through." In particular, older community members have witnessed the climate changing and want to talk about the justice issues they have experienced.

People not facing environmental justice issues might think of justice as a more abstract or philosophical concept. However, those facing it know it is a political struggle. Those of us not dealing with injustices must give attention to these communities and issues to push policy change that will reduce and prevent the environmental hazards these communities face.

Moreover, we must continue to think about framing when attempting to fight injustice. Holloman recommends using the ideal of solidarity to help communities join forces to fight against injustice. No one should be in the fight alone, and helping people work together can be a motivating force and make a greater difference. "If you're speaking to an impacted community, depending on what their particular struggle or struggles are, they'll see it differently and maybe they want to take political action and maybe that'll invigorate them now that they have the terminology and the science behind them.... Keeping it under this umbrella of solidarity gives you more faith," explained Holloman.

Environmental justice is not a standalone issue but often one about communities joining with the goal of overcoming several injustices they face. Since people share the same environment and social concerns, having significant numbers provides power in promoting change.

"The key to making or writing a good story is something that really touches on the human experience. A way to do that is to make sure you're not 'othering' anyone and that you're keen on making connections between the reader and whomever you're writing about and making sure your reader can be empathetic to that person, not as a subject, but as an actual living, breathing human being," said Holloman.

She said we should "focus on things that we as humans have in common. Yes, it is important [to mention or frame with race or socioeconomic details], but you don't have to lead with that. Those can be details later. You can lead with talking about someone and their kids. You can lead with talking about how long this older person has been tilling this land. You can lead with these emblematic stories, or framings that touch at what it is to be *in their shoes*, and then you can talk about some of the potentially polarizing details, but you don't start off with that if the goal is to tell a truly compelling story."

Highlighting stories and not polarization can draw people in to understand environmental justice issues. Therefore, stories about environmental justice should be integrated with mainstream stories about environmental issues.

I also talked with climate communications expert Susan Joy Hassol about ways to be inclusive when discussing environmental issues that have inequitable effects.

"It's really up to all of us to seek out and promote diverse voices. People need to understand that, for example, communities of color are in many ways on the frontlines in climate change. They're more likely to live near power plants and oil refineries, to be exposed to polluted air, to suffer from the many consequences of the fossil fuel economy. So it's really important to hear from those communities and from those voices and to recognize that these effects are not distributed equally. And that in many ways, those least responsible, because they use less energy, are the most and soonest to be affected," explained Hassol, further displaying the need to communicate about environmental justice issues.

"For example, my colleague Michael Mann, a renowned climate scientist, speaks to reporters every day," continued Hassol. She explained that since he can't talk with every reporter, he makes recommendations of other people they can contact, and he promotes diverse voices. "He often recommends women and people of color. So he really makes an effort, and this is something that each of us can do to promote those voices."

The coronavirus further exemplifies an environmental and racial justice issue. "If you're exposed to more air pollution, you're more likely to contract the coronavirus and to suffer a bad outcome from it, and communities of color and poor communities are more likely to be exposed to poor air quality, and more likely to get this terrible virus. So it's really important to seek out and promote those voices and those stories," expounded Hassol.

Social media helps bring more diverse voices into conversations about environmental justice. People facing injustices may use social media to share their stories, and news outlets could pick up on stories and bring more attention to issues. Attention does not solve issues, but it's a necessary step toward justice.

As part of my climate change and psychology class, I did a group project with UNC–Chapel Hill students Rachel Maunus and Claire Bradley. We talked with kids ages eight to twelve to see what they understood about climate justice. We only talked with a few kids, so we didn't have a representative sample, but these kids had a decent grasp on climate science, yet little awareness that climate change affects certain groups of people more than others. We created a podcast to share our experience talking with the kids. I accredit the title of this chapter to Maunus, who named our podcast *It's Not JUST Climate Change*. Environmental justice is not a topic I was taught or read much about when I learned about climate change as a kid; since coming to college, I have better understood that it is an integral part of solving climate change and its related issues.

The Sunrise Movement is one powerful and growing environmental organization that focuses on intersectionality, or the idea that many factors, like class, race, and gender, overlap related to discrimination.[87] As individuals, we can't speak for the experiences

---

87   Merriam-Webster, s.v. "intersectionality (*n.*)," accessed July 5, 2020.

of other people, but we can use our positions and platforms (even if talking to another individual) to remember to discuss justice issues when communicating about the environment. I was able to attend an orientation training with the Sunrise Movement that focused on the organization's strategy for achieving climate justice. I learned the importance of listening to people's stories, particularly related to the environment, without judging but with a focus on understanding and giving them space to share.

The environmental justice organization WE ACT lists specific ways for organizations to fight environmental justice:

1. Provide technical assistance when needed, like legal expertise or trainings on how to conduct water surveys, install rooftop solar panels, or grow community gardens.

2. Create spaces to share resources and network, such as by hosting forums.

3. Evaluate who develops and who benefits from community projects—communities facing problems should be directly involved in creating solutions.

4. Get advice from communities and environmental justice advocates.

5. Intentionally promote diversity and fight oppression like hiring diverse voices and having an advisory board to look for and respond to instances of oppression.

6. Ensure that funds are put toward environment work that benefits marginalized communities.

7. Use your platform to support frontline communities, such as by releasing a statement of solidarity.

8. Make environmental justice part of your mission. Add to your mission statement that you support hiring people with diverse backgrounds and emphasize environmental justice work.[88]

Framing the environment as a justice issue is impactful not just for empowering communities, but also for growing awareness and acceptance of climate change as a problem.

Pope Francis started to accept climate science and advocate for action to reduce climate change when he learned that the climate affects different groups unequally. Understanding that climate change is an ethical issue and believing that caring for and helping those who have less power is important, he became a part of the environmental movement.[89]

The environment is a vast topic, and we can be intentional about the ways we communicate to stop environmental justice from falling through the cracks. We don't just want to stop climate change and return to the status quo. We don't want to preserve the current world, but we should advocate for a better environment that contributes to the well-being of all.

---

88 Morgan Pennington, "8 Ways Environmental Organizations Can Support the Movement for Environmental Justice," WE ACT, accessed July 6, 2020.

89 "Scientists really aren't the best champions of climate science," *Vox*, May 24, 2017.

DISCUSSION QUESTIONS:

- Have you experienced or witnessed environmental injustices and/or environmental racism in your community or one near you? If so, what have people done to fight against these issues, or what could people do?

- Is environmental justice something you see or hear about often when consuming media about climate change? If you have been involved in environmental organizations, was environmental justice emphasized as a major consideration when fighting against climate change? What more could those groups do to fight environmental injustices?

7

SAVE THE EARTH: PROTESTING

—

Skolstrejk för klimatet!

I don't speak Swedish, but this phrase has become very recognizable to me as "school strike for climate." It's a call to action for students across the world to express their deep concerns about the environment. Since Greta Thunberg began her Friday school strikes, climate strikes have been popping up across the globe. Strikes are attention-grabbing and a way to generate newsworthiness to encourage media coverage of looming issues even when climate change and all its associated consequences are not a regular part of the evening news.[90]

I talked to Megan Raisle about her use of a climate rally in Chapel Hill, North Carolina, that served as a visual for communicating climate action needs. I met Raisle when

---

90 Jonathan Watts, "Greta Thunberg, schoolgirl climate change warrior: 'Some people can let things go. I can't,'" *The Guardian*, March 11, 2019.

she was co-chair of the Environmental Affairs Committee in student government my freshman year of college. Raisle is a 2020 graduate of UNC–Chapel Hill, where she studied geography with a minor in environmental studies. She has become more involved in international climate action since representing the university at the United Nations Framework Convention on Climate Change (UNFCCC) Bonn Climate Change Conference.

Raisle helped plan a rally that involved students and Chapel Hill community members biking thirty miles from Chapel Hill to Raleigh to participate in the global climate strike that coincided with the United Nations Climate Change Summit in September 2019. She said the bikes were a good visual to gain attention for the strike, and they were a sustainable form of transportation to reduce carbon dioxide emissions. Ultimately, the goal of the protest was to encourage local and state governments to take climate action.

She said she knew biking far was not accessible for everyone, so she also planned a 1.5-mile bike ride around Chapel Hill. The 1.5 miles were symbolic of the 1.5 degrees Celsius threshold scientists say change needs to remain at to prevent greater climate disaster.

Raisle and others used flyers, chalk messages across campus, and social media to advertise for the event. She was surprised and impressed at the turnout and pleased to see many high school students participating in the climate rally. She noted, "There aren't many issues students are both willing and able to miss class for." The large turnout suggests how meaningful climate action is to young people.

She pointed out that the turnout for the event was not very diverse and most people were white, which reflects the lack of racial diversity in the traditional environmental movement, although people of color have been advocating and dying for their rights for centuries. Raisle's observation further highlights the importance of including diverse groups of people when communicating about the environment.[91]

Ember Penney, a high school student, also organized a strike in Chapel Hill, and she and Raisle ended up working together. Penney said she wanted the Chapel Hill Town Council to create a plan for a Green New Deal. The plan, based on proposed federal legislation with the same name, would make the town more sustainable by reducing fossil fuel use. The federal Green New Deal also focuses on economic growth and increasing jobs.

This event was the first climate strike I have participated in. Remember that feeling of community I mentioned earlier as a way to combat eco-anxiety? Seeing so many adults, college students, and children alike just from the area near my college gather to express their desire for climate action gave me hope. This strike made me remember that none of us are in the fight against climate change alone. In numbers, we can convey how important solving climate change is to so many people.

Ultimately, the strike did not have direct results. The county Chapel Hill is located in began the Orange County Climate Council a week after the protest, but that was planned before

---

91   Diane Toomey, "How Green Groups Became So White and What to Do About It," *Yale Environment 360*, June 21, 2018.

the strike. Additionally, North Carolina has not taken significant climate action. The effects of individual protests can be difficult to measure, but the media attention that Raisle and Penney's protests received show that those protests were well-organized and purposeful.[92]

Strikes tend to make great news stories. Sometimes activism can be impactful and sometimes it can be divisive.

Greta Thunberg has become one of the main faces of the environmental movement among many climate activists. She sends the message that people are angry at businesses and governments for causing climate change. Sometimes this attitude can cause these powerful groups to shift the focus on personal responsibility and generate more political division.[93] Nevertheless, Thunberg has created a global movement.

For example, climate activist Leah Namugerwa of Uganda began weekly school strikes when she was fearful of famine as a result of droughts and landslides due to climate change. She said Thunberg's strikes inspired her to fight despite obstacles like criticism.[94]

Robert Cox, former president of the Sierra Club and a climate rhetoric professor at UNC–Chapel Hill, said Thunberg has

92 Brittany McGee, "Here's what you need to know about the new collaborative Orange County Climate Council," *The Daily Tar Heel*, September 26, 2019.

93 Jake Novak, "How 16-year-old Greta Thunberg's rise could backfire on environmentalists," *CNBC*, September 24, 2019.

94 Inma Gálvez-Robles, "19 Youth Climate Activists You Should Be Following on Social Media," Earth Day Network, June 14, 2019.

been "amazingly effective" and "the global support that she's mobilized" is "now all going on its own, with its momentum with the school strikes movement and the communication networks that have been built around this now." He was not speaking on behalf of the Sierra Club.

From elementary to high school, students are striking and drawing their families' attention to climate change. "That has a ripple effect far broader than just those participating in the strike," said Cox.

The Extinction Rebellion in London uses a different form of protesting. "They engage in civil disobedience that's very disruptive in the middle of city centers, such as blocking the tube in London or highways, because they have come to the conclusion that ordinary protests and marches and the other means of persuasion have all failed. And given that the climate crisis is so dire and happening now that this is their last chance to make a change, and so they're trying to disrupt life as normal in order to get attention. They've had some success in their call for what they call a people's commission to advise the government on steps it can take for climate change," explained Cox. He said other Europeans have also set up citizen commissions to advise governments on taking steps to deal with climate change. "A lot of that has happened as a direct result of the Extinction Rebellion events and all of the media buzz they've gotten." Thus, different environmental groups have different strategies for earning attention, but either way, they have found success.

The main ingredient for a successful protest is strategy. "Protests can outlive their usefulness if they're used too often without being part of a larger strategic effort. I think a lot of people confuse protests as a strategy for a campaign. Protests are simply a tactic, and tactics always have to help implement a larger strategy, and strategy really refers to what's called leverage," explained Cox.

People and groups generally use protests to create attention and display people's desire for change, meaning protests need to be implemented with the goal of generating news coverage in mind. "The protest has to be aligned with a particular occasion, the intended audience for the purpose of the protest, and whether it advances the larger strategy of the group. A lot of protests are used for media gaining efforts.... If a protest outside of the General Assembly in Raleigh attracts enough people, and it's orchestrated in advance with notices out to the media, it will probably get media attention."

Even without the media, protests can help movements in some ways. "You [can] use it as an occasion to mobilize a lot of supporters to give them information for going inside the General Assembly and talking to their individual representatives. I've got a friend who's a Republican in the General Assembly. And he's fairly moderate, and he's a strong environmentalist, but he groans when he hears there's going to be another environmental protest on the street because he doesn't think it affects his Republican colleagues very much. And they turn away," continued Cox.

Since protests are not always the answer for influencing policymakers and affecting change, people can join forces to encourage policy changes in other ways. Joining environmental organizations can provide more numbers and resources to support change. Cox recommended people join "the intervening groups that are large enough to amplify your individual voice in a way that it can be broadcast more."

Again, communications can be effective when we are strategic. "It doesn't cost much to just click or enter your name to a petition. [Politicians] are more responsive if you actually call the office, or even send an old-fashioned letter. They literally count on a daily basis the communications they get that are individual or personal communications, a phone call, a letter of visit, as opposed to just thousands of names on a petition," Cox said.

Climate change is a global issue, so we often think about making changes at the national and international levels. However, people may have more success advocating close to home. "On a local level, I've noticed that public officials do tend to be a little more responsive to public opinion because they are closer to voters, to their citizens or constituents, and they know if they get too far out of line with constituent opinion that they're not going to be reelected," explained Cox.

Communicating about the environment is particularly challenging in the United States. "Compared to [many] other countries, there's so much greater respect for science and simply accepting the science on climate change, and a willingness to take more progressive steps to counter it. For some reason in the United States, there's been a reluctance to accept science.... It's hard to organize a society and move forward without accepting some sources of expertise," said Cox.

Misinformation in the media can provide conflicting information about climate science. But, as you'll learn in the next chapter, media outlets no longer have to give value to climate skepticism.

The public needs to be aware and accepting of environmental problems and the need for policy change. People concerned about the environment must make their voices heard by those in their network and beyond.

DISCUSSION QUESTIONS:

- What have you heard about Thunberg's protests, and how have you reacted to her words and actions?

- Have you ever attended a protest? What was the result? If it was unsuccessful, what might people do to further encourage the changes they hope to see? If you have not attended a protest, think about an environmental issue for which people might protest, and consider how they could be most effective.

8

WORDS OF WORTH
ABOUT EARTH

———

Marine scientist John Bruno received a box of poop after he took a risk by communicating about climate change and overfishing issues.

Unfortunately, some people get upset when professionals, such as scientists and science journalists, advocate for environmental action. Some people think professionals should not show biases or lean to one side or another.

Trying to be unbiased or at least make known our biases is a core tenet of journalism. That being said, communicators do not always need to share "both sides" of an issue when one side is scientific and factual. Communicators should not give attention to untrue anti-environment claims unless they plan to prove them false. Sharing everyone's opinion would be irresponsible.

A science journalist, who asked that her name be withheld because she wasn't authorized to speak on behalf of her employer, explained that when giving proof in journalism, reaching out to scientists for confirmation is no longer necessary if the science is proven. "Debating whether climate change is real rarely needs a quoted expert. When it comes to weather patterns, climate modeling, agricultural impacts, and other newly developing research, it's crucial to speak to experts in the scientific and policy worlds," she said.

Most readers of this book probably have the goal of increasing environmental action, which means you're biased when communicating. We all are. Bias is not necessarily a bad thing, as long as we do not hide it. Even scientists can advocate for change related to the environment these days.

Despite the importance of being strategic when spreading environmental awareness and encouraging action toward sustainability, we must be truthful.

Specific practices around arguments in public relations and journalism can vary. Additionally, scientists face their own set of challenges communicating about the environment and need for action. Ultimately, many professionals have the same goal of describing environmental problems and trying to make the planet a healthier and safer place.

PUBLIC RELATIONS

I talked to Sierra Club Communications Director Maggie Kash to learn more about how an environmental advocacy group communicates strategically but truthfully and to understand strategies used in the public relations aspect of environmental communications.

Kash received a degree from UNC–Chapel Hill's School of Media and Journalism (now the Hussman School of Journalism and Media). She also worked in civil rights advocacy prior to being part of the Executive Team for the Sierra Club.

Kash, who said she was speaking for her own views and not on behalf of the Sierra Club, said the difference between journalism and public relations is journalists tell fair and balanced stories, whereas her role is to give "a compelling narrative for a specific policy or goal."

"Communicating for a nonprofit, you're selling an idea; you're trying to move specific people or audiences," she explained.

She said the mechanics are not very different for environmental advocacy and civil rights advocacy. Both issues are about humanity and how we treat other people. Humanity can draw people into issues before information feels controversial.

Kash said a good example of humanity in media coverage of environmental issues is chronicling the costs—human, economic, and emotional—of climate disasters. She said this coverage is important for demonstrating that climate change is not far off or something that we can wait a decade to solve.

To respond to skepticism and move people toward the Sierra Club's goals as part of the environmental movement, Kash said it's important to "try to meet people where they are and build bridges." She said the Sierra Club does not focus on climate change messages in certain states and regions where people are more skeptical but can instead focus on clean air and water in these regions: "Clean air and water are things I think all people agree they want." Clean air and water are important to the Sierra Club's mission, so focusing on those goals is still worthwhile and more effective than focusing on issues that some people will not be as open to receiving like climate change.

She also told me communications from the Sierra Club are meant to focus on solutions. "We've moved past engaging in a debate about whether or not there is climate change," she said. "Those who wish to debate are generally just trying to distract from the solutions, which we need to be enacting right away."

In my own public relations experience for various environmental organizations, I have also focused on discussing climate change as fact and not a debate. The people who want to read and learn about the environment are unlikely to be skeptics. At this critical point, the most important part is helping people learn how to be part of solutions.

Kash described a "very clear" instance when "the public discourse shifted around climate action." In 2009, then-President Barack Obama gave a speech in the Rose Garden of the White House about his plan for climate action. Kash said up until this point, "'clean coal' was a term tossed by politicians as cover" and they were "pandering to fossil fuel industries." Until then, news channels often used split-screen segments of "fake scientists" debating environmentalists. She said she does not think it is responsible on the media's part "to give that much air time to a fringe group." Kash continued to say that Obama's speech allowed the media, including communications from environmental groups like the Sierra Club, to talk about climate as an issue. Before, climate had to be debated. She said media prior to the speech can be compared

to media now. The wildfires of the late 2010s are linked to climate change in the media, but that link would not have happened ten years earlier.

Tying environmental issues to other issues can broaden audiences and increase support for environmental goals. Anecdotally, Kash has learned that more people have become interested in working at the Sierra Club due to its work with human rights, such as fighting immigration policies.

Kash said her audience depends on the issue. She said in targeting national policy, she has a broader audience. In targeting specific senators, her focus is more regional.

She said people generally have "empathy for animal suffering, and it is a broadly appealing message" to talk about effects of environmental decisions on animals, like the removal of Endangered Species Act protections for gray wolves.

Evidently and not surprisingly, we have multiple ways to draw people into the fight against environmental problems like climate change when specifically encouraging action through public relations.

JOURNALISM

I talked with *Forbes* journalist Jeff McMahon to learn more about ethical environmental journalism.

I asked McMahon, "How do you balance/overcome any personal desire to help the environment while being fair and unbiased in journalism?"

McMahon said, "I adhere to the principle of fairness by including diverse viewpoints and opinions in my stories, especially those of people who are implicated by the stories. However, the question of bias is a tricky one. I don't believe writers can or should be objective—we're human; we're subjective. I think we should be upfront about our views instead, and one of the premises of my work is that the environment matters very much. We cross a line if we begin manipulating, distorting, or misrepresenting the facts in service to any agenda, so it's important to be truthful. As long as we're truthful, we can be honest about our view that the earth is worth protecting."

He said to reach audiences, "There's also an argument that people care more about local impacts than global. I think it's important to be aware of how readers respond and what motivates them and to write in a way that touches upon what they care about, but I also think it's important not to select information to manipulate readers, and also not to assume readers are homogenous. Many do care about land, animals, and global effects." Thus, McMahon can still be strategic in how he reaches audiences to help them best understand environmental problems while being an ethical communicator.

His writing has even been successful in motivating me to take climate action. McMahon wrote an article titled "Greta is Right: Study Shows Individual Lifestyle Change Boosts Systemic Climate Action." I liked this title because it draws interest by mentioning a controversial figure in the environmental movement, Greta Thunberg. It also provides a positive note about climate change that differs from what I often hear and how many people think—that individual action is not impactful—encouraging me to read to learn more.

In his article, he mentioned a significant aspect of reaching audiences and encouraging action: communicators must reduce their own carbon emissions in order to be viewed as credible.[95] So, credibility comes not just from what and how we write, but also from who we are.

Whether writing for journalism or advertising and public relations, communicators should be truthful and use strategy to best help the audience understand a point of view. While these two communications fields differ in their purposes to inform versus persuade, either way communicators must be upfront with their views but also think about the best ways to get those viewpoints across.

When we discuss the environment, people will often want to know supporting information, especially if they are skeptical. Since most of us are not scientists, we can refer to credible sources to support what we say and to guide our audience toward more information.

"Who you reference as a source depends on who you're talking to and who they see as credible," said climate communications expert Susan Joy Hassol. For example, she said NASA and the National Oceanic and Atmospheric Administration (NOAA) are respected sources in climate science that are great quick references in discussions about climate change, and you can also mention sources most relevant to people.

"If you're talking to someone who is a veteran, you could say the US military is really concerned about climate change.

---

95 Jeff McMahon, "Greta Is Right: Study Shows Individual Lifestyle Change Boosts Systemic Climate Action," *Forbes*, November 19, 2019.

They call climate change a threat multiplier, they know it's real and human-caused, and they're very concerned about it as being something that will undermine our security," Hassol told me. This example epitomizes using a source that some may find credible and persuasive about climate change.

Advocates and journalists have their roles in the environmental movement. Where do scientists fit in?

COMMUNICATING AS A SCIENTIST

Let's talk more about Bruno and his will to communicate about the need for environmental action, despite that box of poop someone sent him.

He is a marine sciences professor at UNC–Chapel Hill who studies the impact of climate change on coral reefs and teaches a class about seafood mislabeling. He founded a blog called *Sea Monster* in the 2000s and had an op-ed in *The New York Times* in 2017 about the need to take action to protect coral reefs from climate change.

Bruno said it was questionable in the mid-2000s about whether scientists should engage in climate issues. Experts said scientists should not speak up, and advocates would lose their credibility for displaying bias.

However, since then, scientists advocating for solutions and action has become more appropriate. Stanford climate scientist Stephen Schneider established in 2001 that while some scientists worry about losing their credibility by advocating, they can follow the "responsible scientific advocate model." This model involves scientists recognizing their own biases, clearly sharing factual data and the level of accuracy, and not promoting a sense of fear based on the data.[96] Since then, this model has remained an ethical strategy for scientists. Scientists can set objectives for their communication, like for people to make sustainable decisions in their own lives. Scientists are more effective at using their work for positive social change when they communicate strategically.[97] Philosopher and Oregon State University professor Kathleen Dean Moore says scientists staying silent about the need to take action on climate change is immoral and a human rights violation.[98]

Bruno said he was having dinner with a colleague, and they decided to communicate with the public about environmental issues. They created a blog with the purpose of exciting the public about ocean science and discovery and using it as a vehicle to discuss big threats, like overfishing and climate change.

Bruno has always been interested in fishing and seafood. He wanted to be a chef when he was a kid, and he said he sometimes regrets not going into the restaurant business. He said fishing has historically had a larger impact than climate change on the oceans. For example, fishing has removed 70 to 90 percent of fish biomass from most coral reefs, and most of the world's reefs now lack any top predators. The fishing industry

---

96 Dawn Levy, "Schneider ponders whether scientists should advocate public policy," *Stanford News Service*, May 8, 2001.

97 Anthony Dudo, John Besley, and Shupei Yuan, "Science communication 101: Being strategic isn't unethical," Genetic Literacy Project, December 20, 2017.

98 Jerry Kiesewetter, "Should Scientists Advocate on the Issue of Climate Change?" *Undark*, April 24, 2018.

entails many negative effects. Overfishing and practices that destroy habitats can harm ecosystems and food chains.[99]

He said most people don't know that tuna is analogous to a lion in the ocean, yet many people would be appalled to be served a lion at a restaurant. Therefore, scientists sharing scientific information with the public and using their expertise to help the public understand environment-related topics like ecology are important steps.

Bruno said *Sea Monster* taught him to write quickly, like journalists. He said compared to scientific papers, in journalism "you write shorter, concise, snappy sentences, not formal scientific sense of writing." He also said journalism involves more room for creativity and there is "no filter on blogging," whereas publishing a scientific paper can take two to three years. With the danger of information overload, short writing like journalism and blogging can reach audiences better than scientific papers. Being concise and creative is a way to increase public understanding about complex and dense issues.

However, technology has changed. Blogs are less popular, and social media has taken over. Bruno said Twitter was important for driving people to the blog, but "now Twitter has consumed blogging."

Bruno said when he writes, a lot of the information he wants to share is technical about the way climate change will impact the atmosphere, according to a report from the International

---

99   Jacob Hill, "Environmental Consequences of Fishing Practices," EnvironmentalScience.org, accessed June 14, 2020.

Panel on Climate Change. He said he tries to bring topics back to North Carolina and how it matters to citizens of North Carolina, like hurricanes causing flooding in coastal communities. In other areas, wildfires and drought may be more relevant. He said these issues affect all Americans, from homes flooding to the economy. As we discussed before, the economic effects of waiting to take climate action are extreme.

I asked Bruno how he balances the seriousness of environmental issues with the hope of change.

He said, "The way I balance it—and I don't think of it as hope—I use the word agency or choice. We have a choice, we can control our outcome; it's not black and white; it's not that if we don't stop emissions by 2030, we're locked into a catastrophic future. If they just hear gloom and doom, it's paralyzing."

He tries to show the different lines of choices that can be made. He said humanity does not need to invent anything to solve climate change because we have the necessary technology, but we need to make the choices that will be most sustainable, like using electric cars and energy from windmills or solar panels.

Bruno explained the huge benefits of sustainable choices beyond reducing climate change itself, including global health benefits and high-paying jobs. He said, "We have been so programmed to hear about the costs," but we can enjoy social and ecological benefits too. For example, from working at the East Coast Greenway Alliance, which connects trails from Florida to Maine, I've learned how increasing green space can maintain natural areas and help the economy by encouraging travel and tourism.

Since communicators often need to be sustainable in their own lives to gain credibility, discussing how individuals can be sustainable and what they do is relevant. Bruno explained how individuals can do something about climate change: "My feeling is that there's not a lot that can be done from personal actions. I try to mitigate my emissions as much as I can—riding my bike, stopped eating meat except salmon for protein. I try to limit my flying, although I fly a lot more than most people for my job. We could all do that... [but] fundamental structural changes are needed, which is political. Unfortunately, climate has become politicized, so it's best to vote for politicians that will make decisions to confront it. Personally, you can register voters, raise funds, and run for office."

An issue is people shaming climate scientists for driving and flying, and saying people who do these things are hypocrites. These attacks remind me of when people attack celebrities for supporting the environment. Even if people do not do everything possible to mitigate their environmental impacts, encouraging people to do something and advocate for the environment is better than dissuading people from advocating because they fear being called out.

While people can make changes themselves, they are not easy or significantly impactful, as may be the case for using electric cars, which is why we need effective communications from those who are knowledgeable about it. Although environmental communication experts like Kate Sheppard espouse the importance of trusted messengers, scientists who are engaged in climate work and see the effects firsthand can also share the changes they vouch are important.

Bruno said he has received hate mail for writing about climate for *The New York Times* and thinks this explains why he was sent a box of poop. Despite the greater acceptance of advocacy today, taking a stance on climate change invites intense controversy.

However, he has also received a lot of positive emails thanking him for his work describing climate issues.

Climate change is proven. Although powerful sources like think tanks have made it a partisan issue, climate scientists can and should communicate, not just the data itself, but explanations of it and possible solutions to problems to best reach people and help society combat this major challenge that affects us all.

While not everyone is a scientist, we can similarly share knowledge and stand up for the environment and people who are facing and will face the consequences of irresponsible use of the environment.

DISCUSSION QUESTIONS:
- What do you understand to be the difference between journalism and public relations when communicating about the environment? How can communicators stay credible when discussing environmental action?

- What do you consider to be the place of scientists in advocating about environmental issues society faces?

PART THREE

EFFECTIVE COMMUNICATION EXAMPLES AND LESSONS LEARNED

9

BEYOND PAPER

Art is a lie that makes us realize the truth.

—PABLO PICASSO[100]

Attempting to convince others to take action against climate change can sometimes feel like we're digging into solid rock earth, dry from the increasing droughts from climate change. Words can be powerful, but they can also be tiresome.

Communication isn't just about written words. It's possible—and effective—to use innovative communication methods like art, music, and podcasts to encourage environmental awareness. They can help more people understand the urgency of climate action.

ART
Mary Mattingly combines nature and space to draw audiences. She has created structures reusing materials that raise awareness of environmental issues, which she calls "icons to

---

100 "Famous Pablo Picasso Quotes," PabloPicasso.org, accessed July 1, 2020.

my own consumption." Her work exemplifies an artistic way to communicate about the environment.

Mattingly came to UNC–Chapel Hill to discuss her work with climate. She has had artwork featured in many places, including the Brooklyn Museum.

While describing her efforts to collect the items she used, she said, "I believe materials need to stay in circulation to resonate in the material and subconscious worlds." I interpret this statement as expressing the importance of remembering that the items we use stay around longer than us, so we should not simply use them, throw them away, and forget about them. We should remain conscious of how we are using materials, which will help us lean toward sustainable choices. Mattingly's projects are creative ways to communicate about the circular economy, an idea to be discussed more soon.

Mattingly says two of her famous works are both "publicly occupied" and "stages for storytelling."

One of her projects was the Waterpod, a boat that grows food. Stewards could care for it themselves when they visited the boat. She said it gave people the ability to see a microscopic ecosystem from a macroscopic view. The boat had a direct connection to its environment because the wake from other boats affected the ecosystem. This boat was her home for six months in 2009.

Mattingly grew up in an agricultural town in Connecticut without clean drinking water because of DDT and other chemicals. She came up with the idea to create a boat when the cost of living in New York City became expensive. She wanted to create a "futuristic living space" on the water of New York City. She spent three years getting permits from the Coast Guard and met with the Department of Education about doing workshops and making the space "robust."

She then built the Waterpod entirely from the waste chain in New York City. The wood came from a stage used by the parks department. The soil was city compost. Further, the water used was pumped off rooftops and into tanks.

Another project Mattingly worked on is called Swale. She described it as a "floating edible landscape on a barge" that the community can visit. Mattingly said the boat has had a greater impact on adults, since kids seem to already know to pick food when they arrive to the boat, but adults can be more hesitant. She said seeing soil in the city is difficult, but visiting the boat and witnessing soil impacts can start conversations. Mattingly said people learn about eating locally produced food and discover that doing so is more common than other people think. For example, she said one person talked about picking from a local persimmon tree.

Additionally, Mattingly has piled up her belongings and wrapped them in string to create balls of materials about as tall as her. She said she wanted this project to be "absurd" to show consumption.

Using visual and practical projects like Mattingly's work is a creative way to move from words on pages to tangible ways communities can understand the need to be sustainable and interact with this idea. She sees people conversing about the projects with one another and learning about each other's efforts to be sustainable.

Another example of visual communication is The Plastic Bag store, a model of a grocery store made entirely of plastic. Artist Robin Frohardt said, "I use a lot of humor in my work and I think the subject of plastic and plastic pollution can feel really overwhelming and depressing, but if there's a way to create work that's actually fun and engaging, people might sit with it longer or want to pay attention long enough for some things really to sink in." Part of the humor comes from the names of cereal boxes in the store, like "shredded waste" for "shredded wheats," and "yucky shards" for "lucky charms."[101] Frohardt has displayed the "store" in places from Times Square to Chapel Hill, North Carolina, drawing national attention.[102]

Rob Greenfield is another climate advocate who has drawn attention to human environmental impact through visualization. For one month, he wore all the trash he created around with him through New York City. He also spent a year only eating food he grew and chronicled his journey on social media. His unique practices attracted attention to the idea of living sustainably, and he has gained a large following. Practices like these make being sustainable cool.[103]

---

101 UNC–Chapel Hill, "The Plastic Bag Store," YouTube, September 18, 2018.

102 Caroline Goldstein, "An Artist Is Launching a Pop-Up Grocery Store in Times Square Filled Entirely with Products Made From Upcycled Plastic," Artnet, February 11, 2020.

103 "Rob Greenfield," accessed July 11, 2020.

MUSIC

In addition to art, music can communicate an understanding of environmental changes.

The ClimateMusic Project's goal is for the audience to understand that the climate crisis is a real and urgent issue and to ultimately fight climate change. Music is an intriguing form of climate communications because it can evoke an emotional response that might be harder to get from words on paper. Music can also be more accessible than scientific papers, making it an impactful and useful way to help people understand climate challenges. One of the project's compositions is "Climate."

"Climate" speeds up the tempo to represent an increasing carbon dioxide concentration in the atmosphere. It increases the pitch to represent the rising temperature of earth's atmosphere. The song also uses distortion and volume to represent an imbalance between energy coming to the earth from the sun and being able to leave the earth as heat. Additionally, the compositional form breaks down, representing the pH in the ocean dropping.[104]

When I listened to clips on the website, I felt anxious as the songs sped up, invoking a feeling of urgency in me, so the music achieved its goal. The quick tempo portrayed the fact that the climate is a time-sensitive matter.

The ClimateMusic Project also has a song called "What if We…?" that suggests the joy that could come from climate action—so some music suggests hope, not just despair.[105]

---

104 ClimateMusic, "Our Music," accessed July 6, 2020.

105 Knvul Sheikh, "This Is What Climate Change Sounds Like," *The New York Times*, November 9, 2019.

In addition, Scott St. George, a geography professor at the University of Minnesota, and Daniel Crawford, a music student, created a melody conveying rising global temperatures.[106]

"'Daniel and I have been shocked at how many people continue to contact us because they are moved by the music,' Dr. St. George said. 'When I teach my classes and I put up the latest temperature plots, I don't get that kind of reaction from my students. Graphics just don't land with the same impact,'" shared writer Knvul Sheikh in *The New York Times*.[107] This statement highlights that music is a unique, impactful, and lesser-used way to communicate about climate.

Music on climate change is also increasingly part of mainstream music. The nonprofit environmental news outlet *Grist* called climate anxiety the biggest pop culture trend of 2019. One of the most salient examples is rapper Lil Dicky's video and song called "Earth." It is an anthem spreading care for the planet and its creatures. Its calming and repetitive nature makes it feel like a Vacation Bible School song from my childhood, though it includes some (not necessarily all child-friendly) humor that appeals to Gen Z and continues to make caring about the world cool. The mostly animated video features the voices of many famous singers, including Ariana Grande, Ed Sheeran, Miley Cyrus, and Justin Bieber.[108]

---

106 Ibid

107 Ibid.

108 Zoya Teirstein, "The internet is ablaze with Lil Dicky's bizarre, star-studded climate anthem," *Grist*, April 19, 2019.

More and more young artists are sharing this sentiment of climate change being a threat that looms over our generation. Lil Nas X, singer of "Old Time Road" (the longest song to be #1 on Billboard) said in a tweet that part of the song referred to climate change. Singers Billie Eilish and Lana Del Ray have climate as a "backdrop" to their music.[109] Notably, as we continue to think about reaching audiences, this music may appeal most to young people already most concerned about climate change. Nevertheless, it helps make climate change a more salient issue that people can think about rather than fear.

These innovative and unique forms of communication are indeed impactful. "There's some evidence, according to Anthony Leiserowitz, the director of the Yale Program on Climate Change Communication, that art, and culture more broadly, can shift people into action on climate change. That more artists are addressing it, 'is a mirror to the times,' he said. It's a reflection of our cultural understanding of climate change and also influences our perception of it," wrote Kendra Pierre-Louis and John Schwartz in the Climate Fwd: newsletter in *The New York Times*.[110]

We must convey that environmental issues are serious. But we also must spread that message in ways that will reach people who otherwise might not pay attention. Art and music can help connect with new people as well as provide an outlet for people to engage in environmental communication despite eco-anxiety. I've enjoyed discovering art and music about climate change myself, and I'm excited to see what creators release in the future.

109 Miyo McGinn, "2019's biggest pop-culture trend was climate anxiety," *Grist*, December 27, 2019.

110 Kendra Pierre-Louis and John Schwartz, "Climate Change Burns Its Way Up the Pop Charts," *The New York Times*, May 27, 2020.

PODCASTS

Beyond music, audio is useful for educating and persuading people about the environment and the need to fight climate change.

The digital world has opened up enormous opportunity for reaching audiences and promoting action related to environmental issues. You don't have to be a professional reporter to spread awareness about environmental issues. Student Brooke Bauman developed her own podcast series called *Guilty Plastics*, in which she explored the dependence of society on plastics. Bauman grew up in Chapel Hill, North Carolina, where she enjoyed being part of a jump rope team, before attending UNC–Chapel Hill and majoring in environmental science and getting minors in journalism and geography.

I got to learn more about her podcast on the way to Wrightsville Beach in North Carolina to collect ocean plastics for the Plastic Ocean Project, a nonprofit organization based in Wilmington, North Carolina. The Plastic Ocean Project's "hope is to create a collective community focused on reducing plastic use while finding innovation and collaboration around giving value to plastic waste that will in turn encourage the 'mining' of plastics on and offshore."[111] Ocean plastics cause more than one million seabird deaths and 100,000 marine mammal deaths per year.[112] About 80 percent of ocean plastic comes from land-based sources like agricultural runoff.[113]

111 "About Us—Plastic Ocean Project," Plastic Ocean Project, accessed July 14, 2020.

112 United National Educational, Scientific, and Cultural Organization, "Facts and figures on marine pollution," accessed July 14, 2020.

113 "About Us—Plastic Ocean Project," Plastic Ocean Project, accessed July 14, 2020.

Bauman led a committee that I was part of in Epsilon Eta Environmental Honors Society to help local businesses in Chapel Hill and nearby Carrboro become certified as Ocean-Friendly Establishments by taking steps, such as composting or not using plastic straws, to be more sustainable and less wasteful.

She later explained to me how she got the idea for the podcast. She said she took an audiojournalism class and liked that platform for telling stories because it allowed her to include elements like sounds. Audiojournalism is a way to communicate that includes an authentic human voice, which can have emotion, and allows for multiple people to express their opinions and converse with each other to build points and themes.

Bauman said that at the time she took the class, "I was thinking a lot about plastics and trying to figure out whether individual actions had an impact on such a large-scale problem. It was kind of a perfect storm of those two interests, and [I wanted to create] a podcast exploring this question of whether individual impact [matters]."

To create her podcast, she developed an outline to determine the aspects of plastic she wanted to cover. She said she wanted to include health effects of plastic as well as activism around plastic reduction.

She tried to incorporate scenes she thought would be compelling. She said, "I went to the Plastic Ocean Project, [which] was a good visualization to understand the impact of waste and where it ends up. We were digging around in people's lives [and thinking about how] all their things are gone."

As she dug, Bauman thought about her belongings and wondered, *Where will they go in a hundred years when I'm dead and gone?* She continued, "I think that made me realize our consumption as individuals does matter. And it's hard to say that if every individual changes their consumption patterns [if issues would be solved]. I don't think that's a total part of the solution, but it's a component of it."

Bauman learned about plastic solutions herself through her journey to develop her podcast. She said, "I was thinking beforehand that paper would be better than plastic because paper is biodegradable and composted, but plastic fills landfills and oceans for a long time. But based on carbon footprint research of plastic versus paper, paper is more water intensive and requires a lot of energy to produce." She continued, "We want to make sure that the alternatives are actually better than the stuff you're trying to take away. I think people have expectations going in and [data is important]." Plastic and paper bags both have disadvantages; plastic bags take longer to decompose, and paper bags create more carbon emissions.[114]

Inconclusive answers about which alternative is better, like in this paper versus plastic situation, can complicate communication and shows the need for communication that is not just intriguing but fully explains challenges and benefits of different environmental choices. When people just see short stories and headlines, many might

114 Brad Plumer, "Plastic Bags, or Paper? Here's What to Consider When You Hit the Grocery Store," *The New York Times*, March 29, 2019.

suddenly support getting rid of all plastic bags when that is not necessarily the most sustainable option. Just like with print media, digital media must cover the necessary details and depth.

With the development of easy-to-use podcast apps like Spotify and Podbean, I have seen many people listening to podcasts in recent years, and the ability of a student to create one and communicate about serious issues is inspiring.

The increase of personal technology has allowed many individuals to conduct their own research and create content to share facts and opinions about various science issues. Whether through podcasts or social media, more people can share their thoughts with people across the world.

I love to listen to podcasts because they are convenient. You could listen to them while driving, exercising, crafting, and possibly working, depending on your type of work (e.g., hands-on work). Podcasts are a simple way to acquire more knowledge about what you are interested in or to learn the basics of new ideas and topics.

You can find many environmental podcasts in addition to Bauman's *Guilty Plastics* podcast. Podcasts provide a way for people to share their voices and opinions on niche topics that otherwise would be less available on public radio.

Another environmental podcast is *You Thought You Were An Environmentalist: An Environmental Justice Podcast*. Each issue focuses on a specific environmental justice issue. For example, the podcast chronicles how pesticides for animal agriculture have disproportionate effects on people of color and how Indigenous people face environmental justice issues like the Dakota Access Pipeline on sacred lands. The podcast was created by students at the University of Washington Seattle.[115] Like Bauman, these students used podcasting to spread their voices to the general public.

I also recommend the *Climate Champions* podcast run by Lee Krevat, CEO of Krevat Energy Innovations. Krevat interviews people in a variety of other environmental endeavors, like Scott Anders, director of the Energy Policy Initiatives Center at the University of San Diego.[116] I like this podcast because it provides a positive perspective on the environment by showing action that is being taken, making the future of the environment seem less hopeless while not minimizing the seriousness of the issues.

Another helpful podcast is *Practical(ly) Zero Waste*. Different episodes discuss particular ways for people to reduce their personal waste or the waste in their community. The podcast has an episode about "zero waste on campus" that can appeal to college students, and other episodes on topics like eating locally and buying in bulk.[117]

Full podcasts series are not the only helpful resource; sometimes specific episodes about environmental issues can

115 Julian Barr, *You Thought You Were an Environmentalist: An Environmental Justice Podcast*, Spotify, December 2019.

116 Lee Krevat, "Scott Anders, Director of the Energy Policy Initiatives Center (EPIC), University of San Diego—Episode 15," *The Climate Champions* podcast, Spotify, April 15, 2019.

117 Elsbeth Callaghan, "Zero Waste on Campus," *Practical(ly) Zero Waste* podcast, Spotify, September 1, 2019.

be intriguing. For example, the National Urban League has a podcast called *For the Movement* with an episode about environmental racism issues across the country.[118] Incorporating environmental topics on podcasts not specifically about the environment can draw in more people. This specific podcast is social-justice-oriented, so it is already connected to the environment, and some listeners may already be active environmentalists. Still, having a specific episode dedicated to the environment draws the necessary attention to it.

During my internship with the nonprofit Heart of Waraba, I worked with a team to write and record podcasts about sustainable entrepreneurs across the globe. Creating short episodes was an engaging way to share the entrepreneurs' stories and promote the message that innovation creates hope that we can solve environmental issues.

Although Bauman says making a podcast can be a bit time-consuming and tedious, she thinks it is a rewarding medium. Podcasts are a convenient way to engage in learning about the environment.

I've shared just some of the innovative and informative art, music, and podcasts I have found. I encourage you to search for more creative works and share them with people in your networks, employing strategies we've discussed so far. Think about which of the "Six Americas" they might fall into, from alarmed to dismissive. Think about the language and

---

118 National Urban League, "Environmental Racism: It's A Thing | Flint Mayor Dr. Karen Weaver and Mustafa Ali," *For the Movement* podcast, Spotify, May 28, 2018.

psychology strategies that make sense. Think about including content that highlights all voices and recognizes the justice element of the environment. We can all make change when we share impactful works.

DISCUSSION QUESTIONS:

- What form of innovative communication described in this chapter stood out to you most, and why?

- Brainstorm another creative way to communicate about climate. Think about the audience you want to reach and their current interest and acceptance of climate science. If possible, bring your idea to life!

10

ENGAGEMENT WITH ENVIRONMENTAL ENIGMAS

—

My first almost-all-nighter of college was for my environmental law and policy class. I stayed up late trying to wrap my head around some of the complexities and specifics of the Clean Air Act to prepare for the upcoming final. But I didn't mind. My professor, Dr. Donald Hornstein, taught the class in such an engaging way that I was enthusiastic about understanding technical topics.

Hornstein is an environmental law professor who teaches at UNC–Chapel Hill and Duke University. He has a knack for making complicated legal topics, like the Clean Air Act, and notoriously boring topics, like insurance law, into subjects students love to learn about because of his class. He teaches both undergraduate students not necessarily pursuing law and law students who may not realize aspects of environmental law can be arcane. I met with him to learn his strategies for communicating about these topics and how they can transfer to making the public more engaged with environmental issues.

"The way I teach the undergrad course shares some similarities with the ways I teach at the law school, many of which are technical. You're there to discuss some incredibly obscure regulation that is surrounded by dense administrative law principles," he said.

Some of the tenets Hornstein uses when educating students about environmental law can transfer to educating about the environment in general.

Hornstein employs storytelling to get his points across. "I tell stories about the cases that I know and [these stories] give you the behind-the-scenes information about each of these famous cases and law. The stories I tell are original. I get to decide what stories about the cases I want to tell that work to get students' attention. I use stories in the undergrad class to help animate the statutes, and sometimes even the regs, like the CEQ regs or NEPA." Hornstein is referring to the regulations for the National Environmental Policy Act that explain the process for creating an Environmental Impact Statement. In the class, students apply elements of these statutes to case examples.

Hornstein shares cases as stories, which exemplifies storytelling as a best practice for environmental communication. For example, he explained a case called US *v. Plaza Health* to show the importance of regulating what is put in water. Then he explained the thinking behind the need to implement reasonably achievable technology and the best available control technology as another option to protect water quality. The stories and structure behind specific regulations helped them stick with me. When I visited a wastewater treatment facility in my hometown a year after taking Hornstein's class and learned about the best available technology, I was able to remember the term from class. I can take knowledge from the class and understand how it impacts my own life. I felt pleased to witness good care being taken to clean the water that would be released in streams and sent through the state of North Carolina.

But Hornstein is well aware that his students find the class challenging. "It is in fact a difficult thing to be able to read that level of detail, plug it into arguments, do it under pressure and time, then write it, and embed it in a cogent argument on some larger principle. And so getting students to want to understand these things is a big chunk of why I tell stories and make up stories for that purpose."

Not everyone needs to be a climate scientist or have formal training to understand issues. People can still learn a broad range of issues while receiving enough depth to actually understand them. The strategies Hornstein employs for teaching law to undergrad students can transfer to how people communicate with others about the environment. Communicators should be selective with topics and explain them with enough storytelling, examples, and interaction to keep the audience engaged enough to include details, like how Hornstein is able to use specific sections of laws in his classes.

People often get their news and information from short media bites, social media posts, and articles, making it more rare to get a full understanding of topics. Of course, not everyone can be an expert in everything. Still, sharing the less exciting details in interesting ways can help information resonate with audiences.

"And then I pick my battles on cases," said Hornstein. He said he wanted to reduce information on topics and focus on climate to make the class "a little more current."

We are at a unique time in history, and the particular environmental issues, laws, and available technology are often changing. Hornstein raises the importance of adjusting communications based on current events. He used to spend a lot of time discussing the nonattainment program, which refers to the way the federal government requires and encourages states and localities to reduce their air pollution to certain levels. Hornstein's class explores some of the elements of and changes to this program over time since it was released in the 1970s. He recently started to focus more on game theory, which can apply to countries taking climate action.

"Ironically, [a few weeks into the semester], the Democratic Party released its 600-page climate document of what their climate legislation will look like [if they were in power]," said Hornstein. He must reevaluate the most relevant concepts to teach in environmental law; he may need to focus on this proposed legislation, even though it is not law now.

We should share information that may not be completely certain or in use—in this case, proposed national environmental law. Communicators can help others be aware of possibilities and options when it comes to something as serious as the environment.

Further, Hornstein employs the strategy of appealing to values to engage more people with environmental studies. "I make a point of wanting to make sure people appreciate property rights, and that's linked to freedom. That is not how most students go into thinking about this; it absolutely isn't how most environmentalists look at this. They've used property rights as the enemy," he explained. Conservative students tell him they appreciate getting to learn about property rights and that this concept "opens them up to the rest of the class."

Discussing and supporting property rights when talking about the environment can be a way to cross the partisan divide and draw people with a more conservative mindset to learn about it.

I was surprised to learn how closely connected takings and property law were to environmental law. Normally when I thought of environmental law, I pictured the natural environment and not as much the intersection with ownership of that land, so the course opened my eyes to those connections.

The content Hornstein taught surrounding the Clean Water Act and Clean Air Act has been relevant to my life. I understand them enough to feel knowledgeable when further reading about and working with environmental law and policy. I have seen news stories about changes in water regulations that I can understand. His class also helped me when I began an internship at the EPA's Office of Air Quality Planning and Standards just a few weeks after learning about the structure and elements of the Clean Air Act.

Hornstein continued, "I'm not a popularizer of environmental law." For Hornstein, his goal is not to make more people passionate about the law or to turn them into activists but rather to help people understand law and its relevance.

He intrigues people in topics beyond environmental law. He gave a TEDx Talk on insurance law. "If you think environmental law is arcane, insurance is viewed as the most obscure.... No one comes to law school to learn insurance law," he noted. Yet he was able to excite middle school students about insurance by explaining the relevance of risk in society today, especially as climate change threatens homes on the coast.

I can vouch that Hornstein made me consider insurance an important topic. What I once thought was dry due to stereotypes then seemed fascinating and relevant. He made me consider the consequences of climate change such as sea level rise and natural disasters affecting insurance companies. He made me think about the need to prepare for that and decide how to cover people and whether people are liable if they build homes in areas that are known likely areas to be impacted by climate change. His discussion of coastal impacts also made me think about environmental justice issues that could come from people who have no choice but to live in certain areas that may be affected by climate change.

Hornstein connects insurance to something many people are already interested in—climate—to make people more engaged in the topic of insurance. Similarly, people can communicate about the environment by connecting what others find interesting and important, since not everyone already considers environmental problems to be that (yes, such a news flash at this point, I know). His method is a good example of the strategy to connect environmental problems with people's values.

Further, Hornstein uses humor as a strategy. "I use laughter, all the time. First of all, it makes it fun. I can make points with humor better than I can with anything else."

Satirical humor has been proven effective for changing opinions on serious scientific topics like climate change. Humor on popular shows like *The Late Show* and *Last Week Tonight* can expose more people to science. An example is how more "formerly apathetic" people believed in climate change and considered it a serious problem after watching a satirical video about it from *The Onion*. Viewers also became more concerned about climate change after watching a segment about it on *Jimmy Kimmel Live!* Similar findings happened when satire was used for communicating about controversial topics like vaccines and gene editing.[119]

Researchers found that satirical humor is less likely to receive the backlash typical of science communication and most affects the opinions of least educated viewers.[120]

Even though we should target audiences when trying to influence them, at times we need to consider all audiences. When I interned at the EPA, I witnessed the importance of tailoring communications to a lay audience because the government needs to help everyone know how to stay safe, like avoiding

---

119 Paul R. Brewer and Jessica McKnight, "How Satire Helps Science," *National Geographic*, June 2020. Vol. 237 No. 6.

120 Ibid.

being outdoors on days with poor air quality. I hear about the importance of science communications all the time since a stereotype prevails that scientists are not good at explaining their work. Policy is like science because it is complex, but Hornstein successfully captivates audiences.

I also interviewed a science journalist to learn more about engaging audiences on scientific topics. (She asked me not to share her name as she was not cleared to speak by her employer.) Her writing is a successful example of intriguing audiences to learn more about the environment.

She said to gain attention, you need to find personality in your writing, especially when discussing potentially tricky subjects like energy. She writes as if she were talking to her friend or sister. She says that audiences and journalists are all human, and you should write with that idea in mind. In her work, she is writing to educate and entertain.

"I'm lucky that we have a really playful, bantery audience; I've always vibed with that well. I enjoy finding the fun and being able to find an upbeat voice," said the journalist.

She said a balance exists between knowing and not over-simplifying, and that she strives to be accurate and under-standable. In particular, she said hard science and numbers are difficult to communicate, and scientific jargon certainly doesn't help.

She knows people face different climate change impacts, so she thinks about different angles for her stories. She said environmental issues "are foreign and scary when you don't break them down into understandable bites.... Nowadays knowing a little about everything is important, and the environment affects so many things [like] food, energy, climate, and pollution."

When it comes to discussing complex topics, what is most important is considering your audience and what they know.[121] If they are reading your work, they are probably already interested and somewhat knowledgeable about the topic. If you are reaching out to someone else to spread information, they may be less knowledgeable. Once you have determined your audience, cut down to just the information they need to understand. You don't want to overwhelm your audience with details or complexities.

We also can use metaphors and analogies to provide a better frame of reference to help audience members understand your topic.[122] Climate communications expert Susan Joy Hassol uses a metaphor of a highway to explain why acting now to reduce climate change even while it is already happening is so important. "The best time to have taken action on climate change was twenty-five years ago. But the second best time is now. We have to do it, and we can't go backwards. We need to do as much as we can as fast as we can. If you're driving down a highway, and you miss your exit, what do you do? You slow down, and you take the next exit." She explained that even if we've exceeded certain targets for slowing down climate change, we need to focus on those emissions reductions that

---

121 Ian Altman, "3 Ways Great Speakers Simplify Complex Subjects (without Oversimplifying)," *Inc.*, June 13, 2018.

122 Ibid.

we can achieve. The metaphor has stuck with me when I think about maintaining hope that we have some control over the future of our earth.

Beyond the technicalities of communicating science, we must ensure it resonates. People are unlikely to be compelled to act based on fact. The Alan Alda Center for Communicating Science promotes scientists portraying warmth, benevolence, competency, shared values, and a willingness to listen to most effectively communicate science. Once again, the main principle is connecting with your audience. Leaving time for discussion also helps people engage with materials themselves. Further, we can focus on what is simple to remember. When simplifying the information included, you'll want to balance providing tips that are feasible but also impactful, which are not always the same.[123]

Hornstein wrapped up his thoughts about communicating on complex climate change science and policy by saying, "People need to understand science because it is under attack in a way it hasn't been, maybe ever. This is an attack on science that seems sustained and dangerous. There are forces trying to reverse the course of humanity with these attacks on science."

The details of environmental problems and solutions can be both depressing and complex.

A lot of people will not want to listen to detailed information about environmental science and environmental law,

---

123 Alan Alda Center for Communicating Science, "Making Science Communication More Strategic," YouTube, June 5, 2019.

although a solid understanding of these topics is necessary for taking action toward climate change. People from skeptics to those with eco-anxiety could benefit from entertaining communication to pay attention and feel motivated to act.

If we don't reach our audience, we are wasting our words. That's why we need to keep communications interesting, such as through stories and humor.

DISCUSSION QUESTIONS:
- Think back to a time when you felt engaged in a complicated subject, such as when taking a class or listening to a presentation. What made you feel interested?

- What environmental issues do you still find confusing or even boring to learn about? Brainstorm ways to encourage people to listen and learn about it so they better understand.

11

THE CIRCULAR ECONOMY

———

Let's talk trash.

Food waste has a significant impact on the environment.

**30-40% of food is wasted
in the United States**

People in the United States waste 30 to 40 percent of the food supply.[124]

———

124 "Food Waste FAQS," USDA, accessed July 9, 2020.

I was shocked to learn in a policy class that more than seventy billion pounds of food per year is sent to landfills and combustion facilities in the United States.[125] One ton of food produces 3.8 tons of greenhouse gas emissions, which adds up to many unnecessary greenhouse gas emissions when food is wasted.[126]

Food scarcity is also a problem, but the massive amount of food waste means we should focus on reallocating food, not growing more.[127] Being more intentional with all parts of the supply chain for creating food can reduce its impacts on the environment and contribute to change. Similarly, the constant production of many items like plastics also accumulates to great environmental consequences, like emissions that contribute to climate change.

In addition to food waste, we face the challenge of plastic when transitioning to a more sustainable world. The United States had more than fourteen million tons of plastic in municipal solid waste in 2017.[128]

Ultimately, systematic economic changes will be necessary to reduce the impact of waste on the environment, but individual action, like composting, is necessary to affect

---

125 "How We Fight Food Waste in the US," Feeding America, July 9, 2020.

126 Lindsay Eubanks, "From a Culture of Food Waste to a Culture of Food Security: A Comparison of Food Waste Law and Policy in France and the United States," *William and Mary Environmental Law and Policy Review* 43, no. 2 (2019): 667-687.

127 Karl Deily, "Dispelling Three Food Waste Myths to Protect Our Food Supply," World Food Program USA, October 12, 2017.

128 US EPA, "Plastics: Material-Specific Data," last updated October 30, 2019.

these changes. Media framing is a way to get people to alter their actions.[129]

The circular economy describes how resources can be distributed to reduce unnecessary waste and negative environmental effects of constantly producing goods.

A conference at UNC-Chapel Hill brought attention to the circular economy, a modern way of thinking and communicating about sustainability. I talked to Allie Omens, a 2020 graduate of UNC-Chapel Hill and the president of Epsilon Eta Environmental Honors Society at the time, about her experiences communicating about environmental issues, particularly waste. An environmental studies and public policy major, she led the conference about the circular economy at the university.

She learned about the circular economy in depth when she studied abroad in Copenhagen, Denmark: "Through that, I became really interested in the circular economy as a positive framework of communicating about waste." Omens considers the circular economy a way to innovate out of the waste that we produce. "I came back to [UNC-Chapel Hill] and realized that... a lot of the people I spent time with in the environmental field at Carolina had no idea what the heck I was talking about. It's no one's fault; I came back and realized [the circular economy] wasn't accepted knowledge like it was in Europe."

---

129 Summer Shelton, "Waste not, want not: A media framing evaluation of municipal composting in San Francisco: A city's attempt to combat food waste," Paper presented at International Environmental Communication Association Conference on Communication and the Environment, Leicester, UK (2017).

I had not heard the term "circular economy" until Omens started a committee to host a conference. After the conference, I suddenly heard many people use the term when working on environmental projects. To me, the term brings a revolutionary feel to projects focused on reducing waste, a topic that many people might find unappealing. At the conference, one example of an effort toward a circular economy that stood out to me was Grounded, a business founded by NYU students that transforms coffee grounds into cosmetics.[130] I saw how bringing people together to talk about the environment can be impactful. I learned that we can find solutions to waste, even though it feels like we're locked in a linear pattern of production, consumption, and waste.

Omens also led an initiative to reduce waste in housing at the university. Called Carolina Thrift, the organization collects hundreds of items people no longer want at the end of the school year. The items are then sold at the beginning of the next school year to raise money for sustainable projects. This initiative is one example of how the idea of a circular economy can be implemented at a local or university level.

The "butterfly diagram" is a visual for the circular economy. One side depicts the biological cycles and the other shows the material cycles of resources. The diagram circles because materials should not follow a linear path of being produced, used, and thrown away. Since most people learn about biological cycles like anaerobic digestion and composting, understanding this new way of thinking is digestible; it just adds

---

130 Arin Garland, "Two NYU Students Combine Coffee and Cosmetics," *Washington Square News*, April 8, 2019.

a spin to connect the natural environment with the built environment. The butterfly diagram helps people visualize the need for change and shows that cycling materials is a solution to wasteful consumption.

The circular economy provides a vision for how sustainability can be implemented and viable. Sustainability can be a nebulous idea that makes some people think of radical changes in how we live, but a tangible structure like a butterfly diagram can show steps that reduce waste and emissions.

Thinking about waste reminds me of packing for vacation. I normally think about everything I need to bring, but the waste I will create is less evident and not something I often think to prepare for. When I spent a few days at my grandparents' house, I realized I had not thought about a place to keep my dirty clothes so they would not get mixed in with the clean clothes I packed. I only thought about having clothes and not what I'd do after using them. On a much larger scale, we often do not think about the waste we are creating, because the focus is on production and consumption. What we do with materials is an afterthought.

The World Resources Institute explains that governments can use a circularity hierarchy rather than waste hierarchy. This circularity hierarchy focuses on the use of resources rather than waste management. It limits the materials put into circulation and uses each to its full extent rather than sending it to a landfill or recycling.[131] Even though economic factors are often viewed

---

131 Mathy Stanislaus, "5 Ways to Unlock the Value of the Circular Economy," World Resources Institute, April 15, 2019.

as a block for environmental action, the circular economy is an economically viable way to reduce human impact on the environment. Although the circular economy itself will not stop climate change, it is a necessary step.

Omens discussed the need for strategic communication around issues like the circular economy. "The way we frame environmental issues for non-environmentalists is important. Some of my work experience has shown me how to do the economic argument for environment and for circular economies." For example, she will remind people that throwing away materials is throwing away resources and money. Omens said then people will realize, "That makes sense. I don't need to care about the trees or the whales—this person's not trying to get me not to use straws; they're just trying to explain to me why I shouldn't be throwing away money.'"

From her background with the circular economy, Omens recognizes another unique aspect of environmental communications. She said, "Environmental communication is interesting when you think about the fact that you're speaking on behalf of natural systems that don't have a voice for themselves.... To get people to change their minds, you have to give voices to people who don't have voices." But she points out that people, especially historically oppressed groups, are affected by environmental problems. Environmental quality and people are closely connected.

Omens has also seen issues with environmental communications, especially around climate and waste, when she lived in San Diego. The sea level is rising there, but she does not see residents or local government taking enough environmental action to prevent or adapt to the issue. She said people seem to talk about the need "to act on climate right now" more than actually do something about climate change, like starting a compost, which also contributes to sustainable waste management. Composting is an integral aspect of a circular economy. Instead of going to a landfill, food waste can be used for fertilizers and bio-waste products that can reduce the need for fossil fuels.[132] Yet San Diego, like many places, does not have a compost pickup system.

Making personal sustainable decisions is tough when the infrastructure is not there to make it convenient. When I'm at college, composting is easy for me because my campus has compost bins. On the other hand, composting at home would involve maintaining a compost pile in my yard, paying a company to pick up compost, or bringing it to a place that will accept it. Most municipal areas do not pick up compost the way they might pick up trash and recycling.

I set up a compost and garden at Apex Friendship High School in North Carolina when I founded and led the Environmental Club. The goal was to encourage eating locally—food like onions grown right on campus—and getting rid of extra food sustainably by composting it. But it takes a lot for small groups like clubs to maintain these, and they have only so much space to store compost on a high school campus. Also difficult is explaining these projects to students at a large high school. Lauren Wallace, who led the club after I left, said the group faced the challenge of people contaminating the

---

132 European Compost Network, "Biowaste in the Circular Economy," accessed July 1, 2020.

compost. Ultimately, the club was able to develop a decent compost, though it was still too contaminated to use in the garden. I've also learned from my work in student government that the fear of contamination prevents expansion of composting promotion and access on my university's campus. A policy that widens access to composting along with better education about it could increase the participation in and success of composting.

Omens continued, "I think a lot of people want to do more; they just don't know how." She finds her parents to be a great example of that. Her mom wants to start a compost but does not have clear guidance on how to do that. Her dad wanted to recycle construction materials but could not find a place the government would collect them and had to reuse the materials for another project. She hopes San Diego does more to implement a circular economy.

Creating the change we need and want to see involves many people participating. A great example is San Francisco, which encouraged participation in its compost program by clearly conveying its goal of zero food waste to the public and the need for everyone to contribute, along with fines for excessive food waste. The campaign framed composting as a way for individual citizens to play a significant role in moving toward sustainability. Ultimately, the city decreased greenhouse gas emissions by 12 percent by reducing food waste after this campaign.[133]

---

133 Summer Shelton, "Waste not, want not: A media framing evaluation of municipal composting in San Francisco: A city's attempt to combat food waste," Paper presented at International Environmental Communication Association Conference on Communication and the Environment, Leicester, UK (2017).

Omens also tried to solve an environmental challenge at UNC-Chapel Hill through communication. Many student environmental groups have their own goals and projects and do not collaborate as much as could be helpful. At the same time, students wanted to better understand the university's plan to reduce its environmental impact, called the Three Zeros Plan (meaning net zero water usage, zero waste to landfills, and net zero greenhouse gas emissions). To help students communicate with each other and those at the university involved in the Three Zeros Plan, Omens planned a gathering she called Environment @ Carolina. She had representatives from many environmental organizations, as well as UNC-Chapel Hill faculty involved in the Three Zeros plan, attend.

I attended the meeting myself, and much discussion took place that could have gone on longer than the couple hours allotted. Even though more discussion is needed to reach many students' goals, the meeting served its purpose of providing the opportunity for people with similar and different environmental interests to connect with each other, and the door was open for future communication. I was pleased that some people said they learned about the other aspects of the environmental scene at UNC-Chapel Hill so that moving forward, students and staff can work together on projects that support the circular economy and related environmental issues like climate action. Change normally requires people to communicate what they want.

Environmental groups' struggle to communicate with non-environmentalists is not the only challenge; environmental groups also struggle to effectively communicate with each other. Following the meeting, I started a social media account to combine communication about environmental events and endeavors across campus to be more transparent and to promote collaboration. The account promotes what people are doing, such as getting rid of plastic bags in the dining hall, to show successes and promote more work toward having a circular economy and sustainable use of materials at the university.

"My whole thing is partnerships. Definitely partnerships and communication are the way forward, and we all have bits of information to share with one another so we can solve these huge problems that we're up against. I want to see government partner with the public sector, partner with NGO sector, partner with the people, partner with corporations. There's no reason for us all to not be working together," continued Omens. Better communicating about not just problems but about viable solutions like the circular economy can allow more people to comprehend the need to actively create a more sustainable world.

Whether referring to collaboration among school environmental groups or different industries and the public supporting a more circular economy, a vision for the future as well as partnerships can drive productive conversations. They can help people come together to allow communities and the world to move toward sustainability.

DISCUSSION QUESTIONS:

- Had you heard of the circular economy before reading this book? How is this concept useful for creating a more sustainable society?

- How can we use partnerships to fight against environmental problems? Do you know any other examples of partnerships that have formed to successfully fight against environmental issues?

12

LESSONS FROM
THE PANDEMIC

———

A stark dichotomy exists between life before and after the coronavirus. I wrote the first draft of this book from November 2019 to February 2020; I was in school most of this time and partaking in the hustle and bustle of daily life. But the world was very different as I began the revisions process in March 2020, just as COVID-19 began to quickly spread in the United States and cause lockdowns in many areas. I barely left my home from the start of the revisions process to publication of the book, which reflects the ways global issues can impact individuals. The pandemic has killed hundreds of thousands of people, shattered the economy, and affected almost everyone in some way or another. By heeding the warnings of scientists, taking quick action, and having widespread commitment, we could have saved lives and prevented this tragedy from becoming so severe.

COVID-19 has foreshadowed the severe devastation that will happen without significant climate action. The pandemic teaches us a few lessons:

1. We must pay attention to science, even if it is predictive. At the outset of the pandemic, public health experts did not know what the total number of cases or level of spread would be, but they did warn that losses could be extreme without early action. Similarly, climate scientists have warned about climate change for decades, certain that the earth is warming but uncertain how quickly it will continue doing so. The rate of global warming is unknown because it is dependent upon our actions; that doesn't mean we should not act because of this uncertainty. We need to do more than encourage people to accept science; people must act. We cannot blame a lack of acceptance for science for a worsening environment. Most people don't deny climate change. Taking significant action requires more energy and is more necessary than convincing a minority of people to accept science. With the coronavirus, individual actions matter more when choices can lead to infections and death. With climate, we need most people, not all people, to understand and act.

2. We must recognize that some communities are affected worse than others and work to change that. Native American, Black, and Latinx communities had the highest hospitalization rates from March to June 2020.[134] Not everyone has the privilege of staying isolated during a pandemic. Similarly, we should not be surprised at this point that climate change has the worst effects on minority groups. Communities that are already disadvantaged have the least lifestyle flexibility and resiliency

---

134 "COVID-19 in Racial and Ethnic Minority Groups," CDC, accessed June 25, 2020.

to tragedies from pandemics to natural disasters to food loss and more. Pollution is a risk factor for the coronavirus, and marginalized communities are already more likely to live near pollution sources, which sadly makes them particularly vulnerable to both coronavirus and climate issues.[135]

3. We must involve everyone and show that individual actions help ourselves and our society. Some people make responsible decisions, like social distancing, during a pandemic because they care about others. Other people are responsible because they care about their own health. People have different motivating factors, so communications should cater to both. Similarly, communications about the environment should remind people how environmental action can help both others and themselves. People do not have to be martyrs by making sustainable decisions only for the good of other people and nature.

What does the pandemic show us we can do at local and individual levels to have an impact on climate change? The CDC tells organizations to share information from credible sources to stop the spread of misinformation. It encourages community groups to devote resources toward minority groups to reduce their level of devastation. It tells individuals to take basic precautions like washing their hands.[136] These recommendations all translate well to environmental action

---

135 Rob Wijnberg, "Why climate change is a pandemic in slow motion (and what that can teach us)," *The Correspondent*, May 6, 2020.

136 "COVID-19 in Racial and Ethnic Minority Groups," CDC, accessed June 25, 2020.

and the need to spread correct information, help communities in need, and make individual sustainable choices.

Both tragedies highlight that society does not prepare well for disasters that could have been predicted. Six biases prevent preparation for disasters: (1) myopia, when people think more about immediate costs than future benefits; (2) amnesia, when people forget and do not learn from past disasters; (3) optimism, when people think future disasters are unlikely; (4) inertia, when people choose the status quo when facing uncertainty; (5) simplification, when people only consider some factors when making risky decisions; and (6) herding, when people trust and do the same as those around them.[137] To overcome these issues, we must think about our own tendencies of what risks we may ignore. We should encourage policy changes to spur more extensive change and better preparation.

Another similarity among these tragedies is that we must remain positive and rely on each other to get through them. Since these are global issues, altruism and connection to others is necessary. Certain tips for getting through the pandemic also apply to climate change. We must (1) pay attention to the heroes who are being selfless to make progress fighting these issues. We must (2) practice mindfulness to stay calm and focused so we can make good decisions. We must (3) show gratitude to those doing the right thing, from social distancing to eating less meat, to encourage them to continue. We must (4) remember we are human

---

137 Jill Suttie, "Why Don't We Prepare Enough for Disasters?" *Greater Good Magazine*, March 10, 2017.

and that mistakes, like forgetting to wear a mask or turn off the lights, will happen. When we keep these tips in mind, we can stay motivated to make decisions in the best interest of society.[138]

People can come together over a common issue and change their daily habits, especially when media constantly discusses the virus and what people can do. Imagine if we were in the situation where virtually all media channels—television, radio, online—talked about climate change all day. Increased—and strategic—communications surrounding climate change could similarly create an urge to act quickly. Like we discussed before, the amount of media coverage is a major factor that shapes public opinion around climate change.[139]

Certainly differences exist between these two issues, like how coronavirus is more immediately devastating, whereas climate change is a more insidious threat. We react more to a direct threat compared to something that may feel far away or abstract. And we need to act: the pandemic reminds us that social problems ultimately come down to combined individual decisions. While we are individually drops in a bucket, our actions can have ripple effects.

The pandemic is tragic. But if anything positive came out of the pandemic, lockdowns and an awful economy have

---

138 Jill Suttie, "How to Keep the Greater Good in Mind During the Coronavirus Outbreak," *Greater Good Magazine*, March 10, 2020.

139 Jason T. Carmichael and Robert J. Brulle, "Elite cues, media coverage, and public concern: an integrated path analysis of public opinion on climate change, 2001–2013," *Environmental Politics* (December 5, 2016).

had some positive environmental impacts. Air pollution, electricity usage, and gasoline sales decreased.[140] However, these changes are not forever, and the pandemic has also caused environmental problems in some ways. COVID-19 has had other indirect consequences. Air pollution, while generally terrible, can make clouds block sunlight, leading to some cooling; thus, less pollution is, in a way, creating more global warming. Less gas usage also means gas is cheaper. Low gas costs mean companies can buy new plastic bottles rather than recycled ones, which leads to recycling facilities being shut down. These are more unfortunate and unforeseen impacts of the pandemic.[141] Further, melting sea ice from climate change could release viruses trapped for tens of thousands of years. These tragedies are not just similar but also interconnected. They are both public health threats that will worsen without responsible action.[142] While the pandemic may have slightly slowed climate change, it indicates that our society is not prepared to handle major world changes that affect safety, public health, and the economy, which underscores the necessity of framing climate change as a complex issue, not just about nature. Let's communicate strategically so we can save lives and mitigate the next major crisis of the century.

---

140 Matt Simon, "How Is the Coronavirus Pandemic Affecting Climate Change?" *Wired*, April 21, 2020.

141 Ibid.

142 Rosie McCall, "Melting Glaciers and Thawing Permafrost Could Release Ancient Viruses Locked Away for Thousands of Years," *Newsweek*, February 6, 2020.

DISCUSSION QUESTIONS

- What have you learned from people's individual actions during a pandemic that can be applied to people's individual actions related to climate change?

- What messages have you found effective or ineffective related to responsible actions during the pandemic? Have you tried to convince someone to act responsibly? What does that tell you about advocating for climate action?

CONCLUSION

———

"Climate change is real. It is happening right now, it is the most urgent threat facing our entire species, and we need to work collectively together and stop procrastinating," said actor Leonardo DiCaprio in his acceptance speech for Best Actor at the 2016 Oscars.[143] DiCaprio is famous for championing climate action. And he's not alone. Not all people believe this problem is real and caused by humans, but climate change is a serious threat.

An overwhelming amount of information is available about climate change, and a lot of materials focus on convincing the public that it's true. But many people are already anxious about the state of the environment and want to help work toward sustainability.

People who are alarmed or concerned about environmental issues may struggle with the perception that individuals can do little to solve these threats. Banding together with others

———

143 Yosola Olorunshola, "Leonardo DiCaprio finally wins Oscar! Uses moment to call for climate change action," *Global Citizen*, February 29, 2016.

can help people cope with feelings of helplessness and gain more power to make a difference or encourage others to do the same. Strategic communication is indispensable for helping other people become more knowledgeable about these issues and capable of acting to prevent them.

We don't have much time to change the ways we are affecting the environment; in fact, we are past the time to make changes. Retroactively fixing the environmental problems humans have created will be difficult and more costly. For example, economically viable carbon capture and sequestration technology is not fully developed or likely to be available in the near future. We must take environmentally friendly actions like reducing greenhouse gas emissions and pollutants to save and create a future for all people.

This book focuses on environmental communication as it ought to be in the present age based on the current political and digital landscapes. We are in a unique time in history with polarization, ubiquitous use of social media, and other previously unimaginable events like the coronavirus.

This book is not comprehensive when it comes to information about environmental communication, but I am sharing my journey to learn about a topic that has the power to greatly affect us and the world. I hope it serves as a basis and inspiration for further efforts to strategically communicate about the environment.

My goal is that, from reading this book, you have developed a passion for talking about the environment and feel more equipped to talk to others about it, whether on a large platform or during individual conversations in your life.

Now, we have to recognize that taking small steps to help the environment, like turning off the lights when we leave the room, is not nearly sufficient for solving growing problems like climate change. Such actions feel trite to me at this point. But communicating things we can do to reduce our anxiety in the meantime while we also learn about and help others understand the need for larger change can be beneficial.

If you want to figure out easy ways to communicate, here are some ideas for sustainability in your own life or to share with others who want to live more sustainably. Whenever you notice yourself creating waste or using energy, just ask yourself if you can find a way to make the action a bit more sustainable.

1. Change your eating habits. Eat food produced locally and reduce your meat consumption; go flexitarian. You don't have to say no to bacon for the rest of your life to help the environment. If you want to and can go vegan or vegetarian, then fantastic! But if that prospect is not exciting to you, you can still make efforts to significantly reduce your carbon footprint and water usage by eating less meat. You could make a specific schedule, like only eating meat on weekends, or decide to generally choose non-meat options.

2. Use shampoo and conditioner bars. I began using these as a more sustainable way to wash my hair. While a bit more expensive than a shampoo or conditioner bottle, they can last much longer. I do caution against ordering products online out of enthusiasm for being sustainable. Try to buy these items when going to the store anyway for a larger trip.

3. Carry reusable plastic products around; this action won't solve the problem of plastic oceans, but every little bit helps. Yes, carrying around a water bottle, straw (and cleaner), and utensils is a little awkward unless you're already carrying a large bag. Nevertheless, the more people who do so, the more we can encourage the practice to become mainstream.

I don't live a zero-waste or carbon-emission-free life, but individual steps can keep us motivated to work toward sustainability. While we need more large-scale change, like more renewable energy and a circular economy, we should still take responsibility as individuals as well.

Think about plastic straws. Advocating to stop using them has become common in the environmental movement, but getting rid of all plastic straws will not save most marine life or stop climate change. However, the issue is something tangible that raises awareness of climate change. Environmental work is challenging but fulfilling to me. The future is uncertain, but we must do something rather than nothing. Coming together can allow us to make impactful progress.

People may disagree whether we are responsible as individuals to act sustainably when decades of consumerism and development have led to the environmental problems we face today and going into the future. Nevertheless, in this moment, forgoing individual action makes it worse for ourselves and future generations. Most people do contribute to environmental issues like climate change; in today's world, especially in developed countries, going without producing any waste or emissions is nearly impossible. Many people have challenges in their lives that make it hard to focus on sustainability. If you are concerned about putting food on the dinner table, whether the food was sourced locally probably won't be your greatest concern.

I don't think we should feel guilty or point fingers when we or others act in ways that are not sustainable. Research shows that pride motivates people more than guilt, so we should focus on what people can do rather than what they can't.[144] Many people reducing their environmental impact is better than some people focusing on having no environmental impact. We should strive to do our personal best to be sustainable while we support more significant changes by educating others and advocating for sustainability.

When it comes to our environment, we face numerous problems—the most conspicuous and threatening being climate change. We need to frame climate change as such: we are at a crossroads where we have the power to change outcomes through our actions, though time is running out to do so. The future is malleable. Further, we must continue to frame climate change as a people issue and a justice issue. We can get artistic with communication forms and make communication entertaining. We should post about the polar bears, but we should act beyond that if we actually want to help them. What we can't do is give up. I, along with many others, don't want the earth to be a place where even more people are food-insecure, where people have to flee their land more and more due to climate impacts, or where the current existence

---

144 Claudia Schneider, Lisa Zaval, Elke U. Weber, and Ezra Markowitz, "The influence of anticipated pride and guilt on pro-environmental decision making," PLOS (November 30, 2017).

of creatures like bumblebees and monarch butterflies is just a fantasy.[145]

We have the capability to stop climate change through policy and technology, like encouraging plant-based diets and switching to renewable energy. As we soar into the future, we are in control. Sadly, our society as a whole does not have the will to implement these changes. I'm not sure what the solution is that will bring us together. What I do know is we must keep talking about the environment. Communication is not a panacea, but it is a bridge from a challenge to action. However, we should not just add to the mix of information already out there. We must ensure our communication is as effective as possible.

We can use the tips from many experts featured in this book to have more productive discussions about climate change. We must move beyond facts; instead, we can use strategies like connecting to people's values, storytelling, understanding personality types, and getting creative, among others. We must be strategic if we want to reach people in a polarized nation and digital world. This way, we can bring many people, from as many different viewpoints and backgrounds as possible, into battle. We have to accept that we will not influence every individual, but we should keep communicating as we fight to solve climate change.

For all it's worth, we need to strategically advocate for our earth.

---

145 Jason Gooljar, "10 ANIMALS THREATENED BY CLIMATE CHANGE," Earth Day Network, February 7, 2019.

ACKNOWLEDGMENTS

Thank you to God for providing me this opportunity to write a book and for the blessings that led to this point.

Thank you to my family for their everlasting support, which has continued through this process. Thank you to my mom, Caryn, for being a good listener and for giving me skills like diligence and organization that have helped me reach my goals. Thank you to my dad, Jim, for giving me a sense of joy and calm. Thank you to my sister, Caitlin, for helping me consider unique perspectives. Thank you to my grandparents John and Joanne Reid for encouraging me to have a passion for learning. Thank you to my grandparents Dave and Lorraine Cassidy for encouraging me to pursue my dreams. Thank you to all my other family members for your love and care throughout this process.

Thank you to my interviewees for taking the time to share your thoughts and expertise. I enjoyed all of our discussions, both those in-person conversations and those held remotely due to location or COVID-19.

Thank you to my dear friends, including those at home in the Apex, North Carolina, area and those I've met at UNC–Chapel Hill, for all of your support. Seeing those around me care for each other and the world is inspiring to me.

Thank you to my teachers and mentors throughout the years for fueling my desire to learn and for helping me get where I am today.

Thank you to the Creator Institute and New Degree Press, especially Eric Koester, Brian Bies, Pea Richelle White, and Al Bagdonas for the guidance, mentorship, and encouragement throughout the daunting process of writing a book.

I want to give a big shout-out to the following people who preordered or donated to *Planet Now* to help turn my dream of publishing a book into a reality:

Chloe Allen	Lorraine and Dave Cassidy
Pauline Baggarly	Lydia Cole
Daniel M. Baker	Candi Caulk
Rosie Baker	Michael Dorgan
Brooke Bauman	Jessica Finkel
Lois Boynton	Evelyn Fram
Claire Bradley	Annalee Gault
Cate Byrne	Hannah Green
Elsie I. Cameron	Diane Guthrie
Sara Elizabeth Carr	Molly Hanna

Jonathan Hardin

Madi Hartigan

Ketal Harvey

Suzanne Hurley

Laura Isham

Michael James

Lisa Kelly-Rouse

Mary S. King

Eric Koester

Michelle, John, Andrew, and Ava Marshall

Iniya Muthukumaren

Ian O'Shea

Allie Omens

Maple Osterbrink

Caroline Polito

Stuart Powell

Caitlin Reid

Caryn and Jim Reid

James Reid

Jane Carol Reid

Jeff, Morgan, and Graham Reid and Leigh Garland

Joanne and John Reid

Joe and Sharon Reid

Tyler, Candy, and Tim Richards

John Rothenberg

Layla Saliba

Lily Schwartz

Marguerite Scott

Justin Séguret

Julia Short

Robert A. Speed

Robert T. Speed

Jennifer Te Vazquez

Sydney Layne Thomas

Jordan Trull

Steve Wall

Nancy Williamson Weavil

Jayne Willard

APPENDIX

———

INTRODUCTION

Bendix, Aria. "8 American cities that could disappear by 2100." *Business Insider*, March 17, 2020. https://www.businessinsider.com/american-cities-disappear-sea-level-rise-2100-2019-3.

Breslow, Jason M. "Robert Brulle: Inside the Climate Change 'Countermovement.'" *Frontline*, PBS, October 23, 2012. https://www.pbs.org/wgbh/frontline/article/robert-brulle-inside-the-climate-change-countermovement/.

"Climate Change: How Do We Know?" NASA. Accessed July 8, 2020. https://climate.nasa.gov/evidence/.

D'Amato, Ilario. "Delaying Climate Action Will Raise Costs 50%: World Bank Report." The Climate Group. May 12, 2015. https://www.theclimategroup.org/news/delaying-climate-action-will-raise-costs-50-world-bank-report.

"Global warming of 1.5 °C." Intergovernmental Panel on Climate Change. 2019, 4-9.

Gore, Al et al. *An Inconvenient Truth*. Hollywood, CA: Paramount, 2006. 96 min.

Griswold, Eliza. "How 'Silent Spring' Ignited the Environmental Movement." *The New York Times*, September 21, 2012. https://www.nytimes.com/2012/09/23/magazine/how-silent-spring-ignited-the-environmental-movement.html.

Holden, Emily. "US produces far more waste and recycles far less of it than other developed countries." *The Guardian*, July 3, 2019. https://www.theguardian.com/us-news/2019/jul/02/us-plastic-waste-recycling.

Holland, Greg and Cindy L. Bruyère. "Recent intense hurricane response to global climate change." *Climate Dynamics 42* (2014): 625. DOI 10.1007/s00382-013-1713-0.

Kaplan, Sarah. "By 2050, there will be more plastic than fish in the world's oceans, study says." *The Washington Post*, January 20, 2016. https://www.washingtonpost.com/news/morning-mix/wp/2016/01/20/by-2050-there-will-be-more-plastic-than-fish-in-the-worlds-oceans-study-says/.

Marlon, Jennifer, Peter Howe, Matto Mildenberger, Anthony Leiserowitz and Xinran Wang. "Yale Climate Opinion Maps 2019." Yale Program on Climate Change Communication. September 17, 2019. https://climatecommunication.yale.edu/visualizations-data/ycom-us/.

Merzdorf, Jessica. "A Drier Future Sets the Stage for More Wildfires." NASA. July 9, 2019. https://climate.nasa.gov/news/2891/a-drier-future-sets-the-stage-for-more-wildfires/.

NOAA. "2019 was 2nd hottest year on record for Earth say NOAA, NASA." NOAA. January 15, 2020. https://www.noaa.gov/news/2019-was-2nd-hottest-year-on-record-for-earth-say-noaa-nasa.

Parker, Laura. "Here's How Much Plastic Trash Is Littering the Earth." *National Geographic*, December 20, 2018. https://www.nationalgeographic.com/news/2017/07/plastic-produced-recycling-waste-ocean-trash-debris-environment/.

Peach, Sara. "Sea Surface Temperatures Drive Hurricane Strength." Yale Climate Connections. August 3, 2016. https://www.yaleclimateconnections.org/2016/08/warmer-oceans-stronger-hurricanes/.

Simmons, Daisy. "A brief guide to the impacts of climate change on food production." Yale Climate Connections. September 18, 2019. https://www.yaleclimateconnections.org/2019/09/a-brief-guide-to-the-impacts-of-climate-change-on-food-production/.

Troyer, Rebecca. "Money magazine: Apex is the No. 1 place to live in the US" *Triangle Business Journal*, August 17, 2015. https://www.bizjournals.com/triangle/news/2015/08/17/money-magazine-apex-nc-no-1-best-place-to-live-usa.html.

"UN Report: Nature's Dangerous Decline 'Unprecedented'; Species Extinction Rates 'Accelerating.'" United Nations. May 6, 2019. https://www.un.org/sustainabledevelopment/blog/2019/05/nature-decline-unprecedented-report/.

"What is Sustainability?" UCLA Health. Accessed July 8, 2020. https://www.uclahealth.org/sustainability/about-us.

"World of Change: Global Temperatures." NASA Earth Observatory. Accessed July 8, 2020. https://earthobservatory.nasa.gov/world-of-change/global-temperatures.

CHAPTER 1

"Jeff McMahon." *Forbes*. Accessed May 18, 2020. https://www.forbes.com/sites/jeffmcmahon/#184f82395819.

Leber, Rebecca. "Scott Pruitt's 'Dirty Dealings' Stir a Campaign to Oust Him From the EPA." *Mother Jones*, accessed March 28, 2018. https://www.motherjones.com/environment/2018/03/scott-pruitts-dirty-dealings-stir-a-campaign-to-oust-him-from-the-epa/.

McGrath, Jane. "How Wave Energy Works." How Stuff Works. July 15, 2008.
https://science.howstuffworks.com/environmental/earth/oceanography/wave-energy1.htm.

Wlezien, Christopher and Stuart Soroka. "Public Opinion and Public Policy."
Oxford Research Encyclopedias (April 2016). 10.1093/acrefore/9780190228637.013.74.

CHAPTER 2

Blake, Aaron. "People move to places that fit their politics. And it's helping
Republicans." *The Washington Post*, June 13, 2014.
https://www.washingtonpost.com/news/the-fix/wp/2014/06/13/people-move-to-
places-that-fit-their-politics-and-its-helping-republicans/.

DeSilver, Drew. "The polarized Congress of today has its roots in the 1970s." Pew
Research Center. June 12, 2014.
https://www.pewresearch.org/fact-tank/2014/06/12/polarized-politics-in-congress-
began-in-the-1970s-and-has-been-getting-worse-ever-since/.

"Donald Trump's Election Did Not Increase Political Polarization." Annenberg
School for Communication at the University of Pennsylvania. October 11, 2019.
https://www.asc.upenn.edu/news-events/news/trump-did-not-increase-polarization.

"Eliza Griswold." *The New Yorker*, accessed July 8, 2020.
https://www.newyorker.com/contributors/eliza-griswold.

Fallows, James. "A Nation of Tribes, and Members of the Tribe." *The Atlantic*,
November 4, 2017.
https://www.theatlantic.com/notes/2017/11/a-nation-of-tribes-and-members-of-the-
tribe/544907/.

Griswold, Eliza. "How 'Silent Spring' Ignited the Environmental Movement." *The
New York Times*, September 21, 2012.
https://www.nytimes.com/2012/09/23/magazine/how-silent-spring-ignited-the-
environmental-movement.html.

Hersh, Eitan. "How many Republicans marry Democrats?" *FiveThirtyEight*. June 28, 2016.
https://fivethirtyeight.com/features/how-many-republicans-marry-democrats/.

"How partisans view each other." Pew Research Center. October 10, 2019.
https://www.people-press.org/2019/10/10/how-partisans-view-each-other/.

Kennedy, Brian and Courtney Johnson. "More Americans see climate change as a
priority, but Democrats are much more concerned than Republicans." Pew Research
Center. February 28, 2020.
https://www.pewresearch.org/fact-tank/2020/02/28/more-americans-see-climate-
change-as-a-priority-but-democrats-are-much-more-concerned-than-republicans/.

Krugman, Paul. "Paul Krugman: Australia shows us the road to hell." *The Salt Lake
Tribune*, January 10, 2020.
https://www.sltrib.com/opinion/commentary/2020/01/10/paul-krugman-australia/.

"Most who identify as Republicans and Democrats view their party connection
in positive terms; partisan leaners more likely to cite negative partisanship." Pew
Research Center. October 4, 2017.
https://www.pewresearch.org/politics/2017/10/05/8-partisan-animosity-personal-
politics-views-of-trump/8_04/.

Packer, George. "A New Report Offers Insights Into Tribalism in the Age of Trump." *The New Yorker,* October 13, 2018, https://www.newyorker.com/news/daily-comment/a-new-report-offers-insights-into-tribalism-in-the-age-of-trump.

Pear, Robert. "Cures Act Gains Bipartisan Support That Eluded Obama Health Care Law." *The New York Times,* December 8, 2016. https://www.nytimes.com/2016/12/08/us/politics/cures-act-health-care-congress.html.

Poppick, Laura. "Twelve Years Ago, the Kyoto Protocol Set the Stage for Global Climate Change Policy." *Smithsonian Magazine,* February 17, 2017. https://www.smithsonianmag.com/science-nature/twelve-years-ago-kyoto-protocol-set-stage-global-climate-change-policy-180962229/.

"The Hidden Tribes of America." Hidden Tribes. October 2018. https://hiddentribes.us/.

"The US Endangered Species Act." World Wildlife Fund. Accessed July 8, 2020. https://www.worldwildlife.org/pages/the-us-endangered-species-act.

Yoder, Kate. "On climate change, younger Republicans now sound like Democrats." *Grist,* September 9, 2019. https://grist.org/article/on-climate-change-younger-republicans-now-sound-like-democrat.

CHAPTER 3

"Average Time Spent on Social Media (Latest 2020 Data)." Broadband Search. Accessed June 14, 2020. https://www.broadbandsearch.net/blog/average-daily-time-on-social-media.

Bramwell, Kris. "TikTok videos spread climate change awareness," *BBC,* August 8, 2019. https://www.bbc.com/news/blogs-trending-49202886.

Corcione, Adryan. "What is Greenwashing?" *Business News Daily,* January 17, 2020. https://www.businessnewsdaily.com/10946-greenwashing. html#:~:text=Greenwashing%20is%20when%20a%20company,services%20from%20 environmentally%20conscious%20brands.

"'Fake News' Isn't Easy to Spot on Facebook, According to New Study." University of Texas at Austin. November 5, 2019. https://news.utexas.edu/2019/11/05/fake-news-isnt-easy-to-spot-on-facebook-according-to-new-study/.

"Social Media Fact Sheet." Pew Research Center. June 12, 2019. https://www.pewresearch.org/internet/fact-sheet/social-media/.

CHAPTER 4

Brough, Aaron and James E.B. Wilkie. "Men Resist Green Behavior as Unmanly." Scientific American." December 26, 2017. https://www.scientificamerican.com/article/men-resist-green-behavior-as-unmanly/.

Fishel, Greg. "How a North Carolina meteorologist abandoned his climate change skepticism." *Columbia Journalism Review,* November 13, 2017. https://www.cjr.org/special_report/climate-change-skeptic-meteorologist.php.

"Global Warming's Six Americas." Yale Program on Climate Change Communication. Accessed July 7, 2020. https://climatecommunication.yale.edu/about/projects/global-warmings-six-americas/.

Kamenetz, Anya. "How To Talk To Kids About Climate Change." *NPR*, October 24, 2019. https://www.npr.org/2019/10/22/772266241/how-to-talk-to-your-kids-about-climate-change.

Lawson, D.F., K.T. Stevenson, M.N. Peterson, et al. "Children can foster climate change concern among their parents." *Nat. Clim. Chang.* 9 (2019): 458–462. https://doi.org/10.1038/s41558-019-0463-3.

McCarthy, Joe. "Toxic Masculinity Is Killing the Planet, Study Finds." *Global Citizen*, May 23, 2018. https://www.globalcitizen.org/en/content/toxic-masculinity-is-killing-the-planet-study-fi-2/.

Thurmond, Mikaya. "Science experiment: How carbon dioxide affects oceans." *WRAL*, January 4, 2020. https://www.wral.com/science-experiment-how-carbon-dioxide-affects-oceans/18867037/.

Waite, Richard, Tim Searchinger, and Janet Ranganathan. "6 Pressing Questions About Beef and Climate Change, Answered." World Resources Institute. April 8, 2019. https://www.wri.org/blog/2019/04/6-pressing-questions-about-beef-and-climate-change-answered.

CHAPTER 5

Carmichael, Jason T. and Robert J. Brulle. "Elite cues, media coverage, and public concern: an integrated path analysis of public opinion on climate change, 2001–2013." *Environmental Politics* (December 5, 2016). DOI: 10.1080/09644016.2016.1263433.

"Climate Change in the American Mind." Yale Program on Climate Change Communication and George Mason University Center for Climate Change Communication. November 2019. https://climatecommunication.yale.edu/wp-content/uploads/2019/12/Climate_Change_American_Mind_November_2019b.pdf.

"How the Enneagram System Works." The Enneagram Institute. July 11, 2020. https://www.enneagraminstitute.com/how-the-enneagram-system-works.

Kenyon, Georgina. "Have you ever felt 'solastalgia'?" *BBC*, November 1, 2015. https://www.bbc.com/future/article/20151030-have-you-ever-felt-solastalgia.

Liz Hamel, Lunna Lopes, Cailey Muñana, and Mollyann Brodie. "The Kaiser Family Foundation/Washington Post Climate Change Survey." Kaiser Family Foundation. November 27, 2019. https://www.kff.org/report-section/the-kaiser-family-foundation-washington-post-climate-change-survey-main-findings/.

"Most Effective Communication Strategies With Various Personalities." HRPersonality. Accessed July 11, 2020. https://www.hrpersonality.com/resources/communication-strategies-for-various-personalities.

"Personality Type Explained." Humanmetrics. Accessed July 11, 2020.
http://www.humanmetrics.com/personality/type.

Scher, Avichai. "'Climate grief': The growing emotional toll of climate change." *NBC News*, December 24, 2018.
https://www.nbcnews.com/health/mental-health/climate-grief-growing-emotional-toll-climate-change-n946751.

Sharot, Tali. "How to motivate yourself to change your behavior." YouTube. October 28, 2014.
https://www.youtube.com/watch?v=xp0O2vi8DX4.

"The Climate Dreams Project." Accessed July 7, 2020.
https://climatedreams.com/.

Weiler, Susan, Jason Keller, and Christina Olex. "Personality type differences between Ph.D. climate researchers and the general public: implications for effective communication." *Climatic Change* (August 27, 2009): 237-241. DOI 10.1007/s10584-011-0205-7.

Yancey-Bragg, N'dea. "Here's why people still take the Myers-Briggs test — even though it might not mean anything." *USA Today*, May 6, 2019.
https://www.usatoday.com/story/news/nation/2019/05/06/myers-briggs-type-indicator-does-not-matter/3635592002/.

Yuan, Lily. "The Enneagram Type One - The Perfectionist." Psychology Junkie. September 25, 2019.
https://www.psychologyjunkie.com/2019/09/25/the-enneagram-type-one-the-perfectionist/.

Yuan, Lily. "Here's How You Communicate, Based on Your Enneagram Type." Psychology Junkie. September 10, 2019.
https://www.psychologyjunkie.com/2019/09/10/heres-how-you-communicate-based-on-your-enneagram-type/.

Yoder, Kate. "Frank Luntz, the GOP's message master, calls for climate action." *Grist*, July 25, 2019.
https://grist.org/article/the-gops-most-famous-messaging-strategist-calls-for-climate-action/.

Zak, Dan. "How should we talk about what's happening to our planet?" *The Washington Post*, August 27, 2019.
https://www.washingtonpost.com/lifestyle/style/how-should-we-talk-about-whats-happening-to-our-planet/2019/08/26/d28c4bcc-b213-11e9-8f6c-7828e68cb15f_story.html.

CHAPTER 6

"Environmental Justice and Environmental Racism." Greenaction for Health and Environmental Justice. Accessed July 5, 2020.
http://greenaction.org/what-is-environmental-justice/#:~:text=Environmental%20Racism,movement's%20response%20to%20environmental%20racism.

Gardiner, Beth. "Unequal Impact: The Deep Links Between Racism and Climate Change." *YaleEnvironment360*. June 9, 2020.
https://e360.yale.edu/features/unequal-impact-the-deep-links-between-inequality-and-climate-change.

McLeod, Saul. "Maslow's Hierarchy of Needs." SimplyPsychology. March 20, 2020. https://www.simplypsychology.org/maslow.html.

Merriam-Webster. s.v. "intersectionality (*n.*)." Accessed July 5, 2020. https://www.merriam-webster.com/dictionary/intersectionality.

Pennington, Morgan. "8 Ways Environmental Organizations Can Support the Movement for Environmental Justice." WE ACT. Accessed July 6, 2020. https://www.weact.org/2016/11/863/.

"Scientists really aren't the best champions of climate science." *Vox*, May 24, 2017. https://www.youtube.com/watch?v=-qfI3DZmmQw.

Skelton, Renee and Vernice Miller. "The Environmental Justice Movement." NRDC. March 17, 2016. https://www.nrdc.org/stories/environmental-justice-movement.

CHAPTER 7

Gálvez-Robles, Inma. "19 Youth Climate Activists You Should Be Following on Social Media." Earth Day Network. June 14, 2019. https://www.earthday.org/19-youth-climate-activists-you-should-follow-on-social-media/.

McGee, Brittany. "Here's what you need to know about the new collaborative Orange County Climate Council." *The Daily Tar Heel*, September 26, 2019. https://www.dailytarheel.com/article/2019/09/climate-council-meeting-0927.

Novak, Jake. "How 16-year-old Greta Thunberg's rise could backfire on environmentalists." *CNBC*, September 24, 2019. https://www.cnbc.com/2019/09/24/how-greta-thunbergs-rise-could-backfire-on-environmentalists.html.

Toomey, Diane. "How Green Groups Became So White and What to Do About It." *Yale Environment 360*, June 21, 2018. https://e360.yale.edu/features/how-green-groups-became-so-white-and-what-to-do-about-it.

Watts, Jonathan. "Greta Thunberg, schoolgirl climate change warrior: 'Some people can let things go. I can't.'" *The Guardian*, March 11, 2019. https://www.theguardian.com/world/2019/mar/11/greta-thunberg-schoolgirl-climate-change-warrior-some-people-can-let-things-go-i-cant.

CHAPTER 8

Dudo, Anthony, John Besley, and Shupei Yuan. "Science communication 101: Being strategic isn't unethical," Genetic Literacy Project, December 20, 2017. https://geneticliteracyproject.org/2017/12/20/science-communication-101-being-strategic-isnt-unethical/.

Hill, Jacob. "Environmental Consequences of Fishing Practices." EnvironmentalScience.org. Accessed June 14, 2020. https://www.environmentalscience.org/environmental-consequences-fishing-practices.

Kiesewetter, Jerry. "Should Scientists Advocate on the Issue of Climate Change?" *Undark*, April 24, 2018. https://undark.org/2018/04/24/dilemma-climate-scientist-advocate/.

Levy, Dawn. "Schneider ponders whether scientists should advocate public policy."
Stanford News Service, May 8, 2001.
https://news.stanford.edu/pr/01/schneider59.html.

McMahon, Jeff. "Greta Is Right: Study Shows Individual Lifestyle Change Boosts
Systemic Climate Action." *Forbes*, November 19, 2019.
https://www.forbes.com/sites/jeffmcmahon/2019/11/19/greta-is-right-study-shows-
individual-climate-action-boosts-systemic-change/#1f9fefab4a54.

CHAPTER 9

"About Us - Plastic Ocean Project." Plastic Ocean Project. Accessed July 14, 2020.
https://www.plasticoceanproject.org/about-us.html.

Barr, Julian. *You Thought You Were an Environmentalist: An Environmental Justice
Podcast.* Spotify. December 2019.
https://open.spotify.com/show/2xRmL794Tk6cZIt63bCsGE.

Callaghan, Elsbeth. "Zero Waste on Campus." *Practical(ly) Zero Waste* podcast.
Spotify. September 1, 2019.
https://open.spotify.com/episode/733luUFBNKVo6qukjiVbg6.

"Facts and figures on marine pollution." United National Educational, Scientific, and
Cultural Organization. Accessed July 14, 2020.
http://www.unesco.org/new/en/natural-sciences/ioc-oceans/focus-areas/rio-20-
ocean/blueprint-for-the-future-we-want/marine-pollution/facts-and-figures-on-
marine-pollution/.

"Famous Pablo Picasso Quotes." PabloPicasso.org. Accessed July 1, 2020.
https://www.pablopicasso.org/quotes.jsp.

Goldstein, Caroline. "An Artist Is Launching a Pop-Up Grocery Store in Times
Square Filled Entirely With Products Made From Upcycled Plastic." Artnet.
February 11, 2020.
https://news.artnet.com/exhibitions/plastic-bag-store-coming-times-
square-1774794.

Krevat, Lee. "Scott Anders, Director of the Energy Policy Initiatives Center (EPIC),
University of San Diego - Episode 15." *The Climate Champions* podcast. Spotify.
April 15, 2019.
https://open.spotify.com/episode/3lBfol8WpJuFmkT6wiDNnh.

McGinn, Miyo. "2019's biggest pop-culture trend was climate anxiety." *Grist*,
December 27, 2019.
https://grist.org/politics/2019s-biggest-pop-culture-trend-was-climate-anxiety/.

National Urban League. "Environmental Racism: It's A Thing | Flint Mayor Dr.
Karen Weaver and Mustafa Ali." *For the Movement* podcast. Spotify. May 28, 2018.
https://open.spotify.com/episode/owDNVyDmVVANOi4pgw5D9W.

"Our Music." ClimateMusic. Accessed July 6, 2020.
https://climatemusic.org/our-music.

Pierre-Louis, Kendra and John Schwartz. "Climate Change Burns Its Way Up the
Pop Charts." *The New York Times*, May 27, 2020.
https://www.nytimes.com/2020/05/27/climate/nyt-climate-newsletter-pop-songs.html.

Plumer, Brad. "Plastic Bags, or Paper? Here's What to Consider When You Hit the Grocery Store." *The New York Times*, March 29, 2019. https://news.artnet.com/exhibitions/plastic-bag-store-coming-times-square-1774794.

"Rob Greenfield." Accessed July 11, 2020. https://www.robgreenfield.org/.

Sheikh, Knvul. "This Is What Climate Change Sounds Like." *The New York Times*, November 9, 2019. https://www.nytimes.com/2019/11/09/science/climate-change-music-sound.html.

Teirstein, Zoya. "The internet is ablaze with Lil Dicky's bizarre, star-studded climate anthem." *Grist*, April 19, 2019. https://grist.org/article/the-internet-is-ablaze-with-lil-dickys-bizarre-star-studded-climate-anthem/.

UNC–Chapel Hill. "The Plastic Bag Store." YouTube. September 18, 2018. https://www.youtube.com/watch?v=hemPPkjH62g.

CHAPTER 10

Alan Alda Center for Communicating Science. "Making Science Communication More Strategic." YouTube. June 5, 2019. https://www.youtube.com/watch?v=C5fqUJcswJQ.

Altman, Ian. "3 Ways Great Speakers Simplify Complex Subjects (Without Oversimplifying)." *Inc.*, June 13, 2018. https://www.inc.com/ian-altman/3-ways-to-explain-complex-topics-simply-without-oversimplifying.html.

Brewer, Paul R. and Jessica McKnight. "How Satire Helps Science." *National Geographic*, June 2020. Vol. 237, no. 6.

CHAPTER 11

"Biowaste in the Circular Economy." European Compost Network. Accessed July 1, 2020. https://www.compostnetwork.info/policy/circular-economy/.

Deily, Karl. "Dispelling Three Food Waste Myths to Protect Our Food Supply." World Food Program USA. October 12, 2017. https://www.wfpusa.org/stories/dispelling-food-waste-myths/.

Eubanks, Lindsay. "From a Culture of Food Waste to a Culture of Food Security: A Comparison of Food Waste Law and Policy in France and the United States." *William and Mary Environmental Law and Policy Review* 43, no 2 (2019): 667-687.

"Food Waste FAQs." USDA. Accessed July 9, 2020. https://www.usda.gov/foodwaste/faqs#:~:text=In%20the%20United%20States%2C%20food,worth%20of%20food%20in%202010.

Garland, Arin. "Two NYU Students Combine Coffee and Cosmetics." *Washington Square News*, April 8, 2019, https://nyunews.com/culture/dining/2019/04/08/nyu-create-sustainable-upcycled-coffee-soap-business-grounded/.

"How We Fight Food Waste in the US." Feeding America. July 9, 2020. https://www.feedingamerica.org/our-work/our-approach/reduce-food-waste#:~:text=72%20billion%20pounds%20of%20food%20is%20lost%20each%20year%2C%20not,landfills%20and%20incinerators%20every%20year.

"Plastics: Material-Specific Data." US EPA. Last updated October 30, 2019. https://www.epa.gov/facts-and-figures-about-materials-waste-and-recycling/plastics-material-specific-data.

Shelton, Summer. "Waste not, want not: A media framing evaluation of municipal composting in San Francisco: A city's attempt to combat food waste." Paper presented at International Environmental Communication Association Conference on Communication and the Environment, Leicester, UK. (2017).

Stanislaus, Mathy. "5 Ways to Unlock the Value of the Circular Economy." World Resources Institute. April 15, 2019. https://www.wri.org/blog/2019/04/5-ways-unlock-value-circular-economy.

CHAPTER 12

Carmichael, Jason T. and Robert J. Brulle. "Elite cues, media coverage, and public concern: an integrated path analysis of public opinion on climate change, 2001–2013." Environmental Politics (December 5, 2016). DOI: 10.1080/09644016.2016.1263433.

"COVID-19 in Racial and Ethnic Minority Groups." CDC. Accessed June 25, 2020. https://www.cdc.gov/coronavirus/2019-ncov/need-extra-precautions/racial-ethnic-minorities.html.

McCall, Rosie. "Melting Glaciers and Thawing Permafrost Could Release Ancient Viruses Locked Away for Thousands of Years." Newsweek, February 6, 2020. https://www.newsweek.com/melting-glaciers-thawing-permafrost-ancient-viruses-1486037.

Simon, Matt. "How Is the Coronavirus Pandemic Affecting Climate Change?" Wired, April 21, 2020. https://www.wired.com/story/coronavirus-pandemic-climate-change/.

Suttie, Jill. "How to Keep the Greater Good in Mind During the Coronavirus Outbreak." Greater Good Magazine, March 10, 2020. https://greatergood.berkeley.edu/article/item/how_to_keep_the_greater_good_in_mind_during_the_coronavirus_outbreak.

Suttie, Jill. "Why Don't We Prepare Enough for Disasters?" Greater Good Magazine, March 10, 2017. https://greatergood.berkeley.edu/article/item/why_dont_we_prepare_enough_for_disasters.

Wijnberg, Rob. "Why climate change is a pandemic in slow motion (and what that can teach us)." The Correspondent, May 6, 2020. https://thecorrespondent.com/449/why-climate-change-is-a-pandemic-in-slow-motion-and-what-that-can-teach-us/10477915635-ffbbde9b.

CONCLUSION

Gooljar, Jason. "10 Animals Threatened by Climate Change." Earth Day Network. February 7, 2019.
https://www.earthday.org/how-climate-change-is-threatening-our-species/.

Olorunshola, Yosola, "Leonardo DiCaprio finally wins Oscar! Uses moment to call for climate change action." *Global Citizen*, February 29, 2016.
https://www.globalcitizen.org/en/content/leonardo-dicaprio-calls-for-urgent-action-against/.

Schneider, Claudia, Lisa Zaval, Elke U. Weber, and Ezra Markowitz. "The influence of anticipated pride and guilt on pro-environmental decision making." PLOS ONE (November 30, 2017).
https://doi.org/10.1371/journal.pone.0188781.